EVERY STUDENT CAN LEARN MATHEMATICS

MATHEMATICS Instruction & Tasks
in a PLC at Work®

Timothy D. Kanold Jessica Kanold-McIntyre
Matthew R. Larson Bill Barnes
Sarah Schuhl Mona Toncheff

Solution Tree | Press

A Joint Publication With

NCTM® NATIONAL COUNCIL OF TEACHERS OF MATHEMATICS

Copyright © 2018 by Solution Tree Press

Materials appearing here are copyrighted. With one exception, all rights are reserved. Readers may reproduce only those pages marked "Reproducible." Otherwise, no part of this book may be reproduced or transmitted in any form or by any means (electronic, photocopying, recording, or otherwise) without prior written permission of the publisher.

555 North Morton Street
Bloomington, IN 47404
800.733.6786 (toll free) / 812.336.7700
FAX: 812.336.7790

email: info@SolutionTree.com
SolutionTree.com

Visit **go.SolutionTree.com/MathematicsatWork** to download the free reproducibles in this book.

Printed in the United States of America

Library of Congress Cataloging-in-Publication Data

Names: Kanold, Timothy D., author.
Title: Mathematics instruction and tasks in a PLC at work / Timothy D. Kanold
  [and five others].
Other titles: Mathematics instruction and tasks in a professional learning
  community at work
Description: Bloomington, IN : Solution Tree Press, [2018] | Includes
  bibliographical references and index.
Identifiers: LCCN 2017046582 | ISBN 9781945349997 (perfect bound)
Subjects: LCSH: Mathematics--Study and teaching. | Mathematics
  teachers--Training of.
Classification: LCC QA11.2 .G7427 2018 | DDC 510.71/2--dc23 LC record available at https://lccn
  .loc.gov/2017046582

Solution Tree
Jeffrey C. Jones, CEO
Edmund M. Ackerman, President

Solution Tree Press
President and Publisher: Douglas M. Rife
Editorial Director: Sarah Payne-Mills
Art Director: Rian Anderson
Managing Production Editor: Caroline Cascio
Senior Production Editor: Suzanne Kraszewski
Senior Editor: Amy Rubenstein
Copy Editor: Ashante K. Thomas
Proofreader: Elisabeth Abrams
Text and Cover Designer: Laura Cox
Editorial Assistants: Jessi Finn and Kendra Slayton

Acknowledgments

Timothy D. Kanold

First and foremost, my thanks to each author on our team for understanding the joy, pain, and hard work of the writing journey, and for giving freely of their deep mathematics talent to so many others. Sarah, Matt, Bill, Mona, and especially my daughter, Jessica, you are each a gifted colleague and friend.

My thanks also to our reviewers, colleagues who have dedicated their lives to the work and effort described within the pages of this assessment book, and especially Kit, Sharon, and Jenn, who agreed to dedicate the time necessary to review each book in the series. I personally am grateful to each reviewer, and I know that you also believe that *every student can learn mathematics.*

Thanks, too, to Jeff Jones, Douglas Rife, Suzanne Kraszewski, and the entire editorial team from Solution Tree for their belief in our vision and work in K–12 mathematics education and for making our writing and ideas so much better.

Thanks also to my wife, Susan, and the members of our "fambam" who understand how to love me and formatively guide me through the good and the tough times a series project as bold as this requires.

And finally, my thanks go to you, the reader. May you rediscover your love for mathematics and teaching each and every day. The story of your life work matters.

Jessica Kanold-McIntyre

I am extremely grateful to the writing team of Bill, Mona, Sarah, Tim, and Matt for their inspiration and collaboration through this writing process. Thank you to the teachers and leaders that have encouraged and influenced me along my own professional journey. Your passion and dedication to students and student learning continue to help me grow and learn. A special thanks to my husband for his support and encouragement throughout the process of writing this book. And, last but not least, I'd like to acknowledge our two dogs, Mac and Lulu, and my daughter Abigail Rose, who was born during the writing of this book, for their constant love.

Matthew R. Larson

It was an honor to be a member of an authorship team that truly understands each and every student can learn mathematics at a high level if the necessary conditions are in place. I thank all of you for your leadership and commitment to students and their teachers.

My thanks also go out to the thousands of dedicated teachers of mathematics who continue their own learning and work tirelessly every day to reach their students to ensure they experience mathematics in ways that prepare them for college and careers, promote active engagement in our democratic society, and help them experience the joy and wonder of learning mathematics. I hope this book will support you in your critical work.

Bill Barnes

First and foremost, thank you to our authorship team, Sarah, Mona, Jessica, and Matt, colleagues and friends who helped me overcome challenges associated with writing about the deep-rooted culture associated with homework and grading practices. Special thanks to Tim Kanold for continuing to believe in me, for supporting my work, and for helping me find my voice.

Special thanks to my wife, Page, and my daughter, Abby, who encourage me to pursue a diverse set of professional and personal interests. I am so grateful for your love, support, and patience as I continue to grow as a husband and father.

Finally, thanks to my work family, friends, and colleagues from the Howard County Public School System. I would like to especially thank our superintendent, Dr. Michael J. Martirano, for insisting that I follow my passion and continue to grow as a professional.

Sarah Schuhl

Working to advance mathematics education within the context of a professional learning community can be a daunting task. Many thanks to Tim for his vision, encouragement, feedback, and support. Since our paths first crossed, he has become my mentor, colleague, and friend. Your heart for students and educators is a beautiful part of all you do.

Additionally, thank you to my dear friends and colleagues Mona, Bill, Jessica, and Matt, led by Tim in his team approach to completing this series. I am grateful to work with such a talented team of educators who are focused on ensuring the mathematics learning of each and every student. You have challenged me and made the work both inspiring and fun.

To the many teachers I have worked with, thank you for your tireless dedication to students and your willingness to let me learn with you. A special thank you to the teachers at Aloha-Huber Park School for working with me to give tasks so we could gather evidence of student reasoning and thinking to share.

Finally, I was only able to be a part of this series because of the support, love, and laughter given to me by my husband, Jon, and our sons, Jacob and Sam. Your patience and encouragement mean the world to me and will be forever appreciated.

Mona Toncheff

During my educational career, I have had the privilege of teaching students with diverse learning needs and backgrounds. Each year, my goal is to make mathematics more accessible to inspire students to continue their mathematics journey.

Now, I have the privilege of working with teachers and leaders across the nation whose goal is the same. The teachers and leaders with whom I have learned over the past twenty-five years have inspired this book series. The vision for a deeper understanding of mathematics for each and every student is achievable when we collectively respond to diverse student learning.

I am forever grateful to my husband, Gordon, and team Toncheff, who always support and encourage my professional pursuits.

Finally, thank you Tim for your leadership, mentorship, and vision for this series; and to Sarah, Matthew, Bill, and Jess; I am a better leader because of learning and leading with you.

Solution Tree Press would like to thank the following reviewers:

Suyi Chuang
Supervisor of Professional Learning
Loudoun County Public Schools
Ashburn, Virginia

Jennifer Deinhart
K–8 Mathematics Specialist
Mason Crest Elementary School
Annandale, Virginia

Jill Gough
Director of Teaching and Learning
Trinity School
Atlanta, Georgia

Nathan Lang
Chief Education Officer, WeVideo
San Francisco, California

Kit Norris
Educational Consultant
Hudson, Massachusetts

Barbara Perez
Director, K–12 Mathematics, Instructional Design,
and Professional Learning Division
Clark County School District
Las Vegas, Nevada

Sharon Rendon
Mathematics Consultant
National Council of Supervisors of Mathematics
Central 2 Regional Director
Rapid City, South Dakota

Cassie Sisemore
Teacher on Special Assignment, Secondary
Mathematics
Instructional Services, Visalia Unified School District
Visalia, California

Visit **go.SolutionTree.com/MathematicsatWork**
to download the free reproducibles in this book.

Table of Contents

About the Authors ix
Preface xiii
Introduction 1
    Equity and PLCs 1
    The Reflect, Refine, and Act Cycle 2
    Team Actions and the Mathematics in a PLC at Work Framework 3
    About This Book 3

PART 1
Team Action 3: Develop High-Quality Mathematics Lessons for Daily Instruction 7
    Your Mathematics Instruction *Purpose* 7
    The Mathematics in a PLC at Work Instructional Framework 8

1 Essential Learning Standards: The *Why* of the Lesson 13
    Developing Clarity 13
    Reflecting on Practice 18

2 Prior-Knowledge Warm-Up Activities 19
    Choosing Prior-Knowledge Warm-Up Activities 19
    Reflecting on Practice 24

3 Academic Language Vocabulary as Part of Instruction 27
    Incorporating Academic Language Vocabulary 28
    Reflecting on Practice 28

4 Lower- and Higher-Level-Cognitive-Demand Mathematical Task Balance 33
    Definition of Higher-Level- and Lower-Level-Cognitive-Demand Tasks 34
    Why Balancing Use of Lower- and Higher-Level-Cognitive-Demand Tasks Is Important 34
    Reflecting on Practice 36

5 Whole-Group and Small-Group Discourse Balance ... 39
Facilitating Discourse in the Classroom ... 39
Reflecting on Practice ... 43

6 Lesson Closure for Evidence of Learning ... 47
Student-Led Summaries ... 47
Sample Closing Prompts ... 49
Reflecting on Practice ... 49

7 Mathematics in a PLC at Work Lesson-Design Tool ... 51
Planning With Your Team ... 51
Reflecting on Practice ... 55

Part 1 Summary ... 57

PART 2
Team Action 4: Use Effective Lesson Design to Provide Formative Feedback and Student Perseverance ... 61

Essential Characteristics of Meaningful FAST Feedback ... 62
Checking for Understanding and the Formative Feedback Process ... 63

8 Essential Learning Standards and Prior-Knowledge Warm-Up Activities ... 67
Guidelines to Consider ... 69

9 Using Vocabulary as Part of Instruction ... 71
When and How to Teach Vocabulary ... 72

10 Implementing Mathematical Task and Discourse Balance ... 77
Support Student Perseverance During Whole-Group Discourse ... 82
Support Student Perseverance During Small-Group Discourse ... 86
Monitor Actions and Results ... 88
Manage Student Teams ... 90

11 Using Lesson Closure for Evidence of Learning ... 99
Facilitating Closure Activities ... 99
Reflecting on Effectiveness ... 102

12 Responding to Lesson Progress With High-Quality Tier 1 Mathematics Intervention ... 107
Planning Your In-Class Interventions ... 107
Analyzing Data for Tier 1 Intervention ... 109

Part 2 Summary ... 113

Epilogue ... 115
Appendix: Cognitive-Demand-Level Task Analysis Guide ... 117
References and Resources ... 120
Index ... 124

About the Authors

Timothy D. Kanold, PhD, is an award-winning educator, author, and consultant and national thought leader in mathematics. He is former director of mathematics and science and served as superintendent of Adlai E. Stevenson High School District 125, a model professional learning community (PLC) district in Lincolnshire, Illinois.

Dr. Kanold is committed to equity and excellence for students, faculty, and school administrators. He conducts highly motivational professional development leadership seminars worldwide with a focus on turning school vision into realized action that creates greater equity for students through the effective delivery of the PLC process by faculty and administrators.

He is a past president of the National Council of Supervisors of Mathematics (NCSM) and coauthor of several best-selling mathematics textbooks over several decades. Dr. Kanold has authored or coauthored thirteen books on K–12 mathematics and school leadership since 2011, including the bestselling book *HEART!* He also has served on writing commissions for the National Council of Teachers of Mathematics (NCTM) and has authored numerous articles and chapters on school leadership and development for education publications since 2006.

Dr. Kanold received the 2017 Ross Taylor/Glenn Gilbert Mathematics Education Leadership Award from the National Council of Supervisors of Mathematics, the international 2010 Damen Award for outstanding contributions to the leadership field of education from Loyola University Chicago, 1986 Presidential Awards for Excellence in Mathematics and Science Teaching, and the 1994 Outstanding Administrator Award from the Illinois State Board of Education. He serves as an adjunct faculty member for the graduate school at Loyola University Chicago.

Dr. Kanold earned a bachelor's degree in education and a master's degree in mathematics from Illinois State University. He also completed a master's degree in educational administration at the University of Illinois and received a doctorate in educational leadership and counseling psychology from Loyola University Chicago.

To learn more about Timothy D. Kanold's work, visit his blog, *Turning Vision Into Action* (www.turningvisionintoaction.today) and follow him on Twitter @tkanold.

Jessica Kanold-McIntyre is an educational consultant and author committed to supporting teacher implementation of rigorous mathematics curriculum and assessment practices blended with research-informed instructional practices. She works with teachers and schools around the country to meet the needs of their students. Specifically, she specializes in building and supporting the collaborative teacher culture through the curriculum, assessment, and instruction cycle.

She has served as a middle school principal, assistant principal, and mathematics teacher and leader. As principal of Aptakisic Junior High School in Buffalo Grove, Illinois, she supported her teachers in implementing initiatives, such as the Illinois Learning Standards; Next Generation Science Standards; and the College, Career, and Civic Life Framework for Social Studies State Standards, while also supporting a one-to-one iPad environment for all students. She focused on teacher instruction through the PLC process, creating learning opportunities around formative assessment practices, data analysis, and student engagement. She previously served as assistant principal at Aptakisic, where she led and supported special education, response to intervention (RTI), and English learner staff through the PLC process.

As a mathematics teacher and leader, Kanold-McIntyre strived to create equitable and rigorous learning opportunities for all students while also providing them with cutting-edge 21st century experiences that engage and challenge them. As a mathematics leader, she developed and implemented a districtwide process for the Common Core State Standards in Illinois and led a collaborative process to create mathematics curriculum guides for K–8 mathematics, algebra 1, and algebra 2. She currently serves as a board member for the National Council of Supervisors of Mathematics (NCSM).

Kanold-McIntyre earned a bachelor's degree in elementary education from Wheaton College and a master's degree in educational administration from Northern Illinois University. To learn more about Jessica Kanold-McIntyre's work, follow her on Twitter @jkanold.

Matthew R. Larson, PhD, is an award-winning educator and author who served as the K–12 mathematics curriculum specialist for Lincoln Public Schools in Nebraska for more than twenty years. He served as president of the National Council of Teachers of Mathematics (NCTM) from 2016–2018. Dr. Larson has taught mathematics at the elementary through college levels and has held an honorary appointment as a visiting associate professor of mathematics education at Teachers College, Columbia University.

He is coauthor of several mathematics textbooks, professional books, and articles on mathematics education, and was a contributing writer on the influential publications *Principles to Actions: Ensuring Mathematical Success for All* (NCTM, 2014) and *Catalyzing Change in High School Mathematics: Initiating Critical Conversations* (NCTM, 2018). A frequent keynote speaker at national meetings, Dr. Larson's humorous presentations are well-known for their application of research findings to practice.

Dr. Larson earned a bachelor's degree and doctorate from the University of Nebraska–Lincoln, where he is an adjunct professor in the department of mathematics.

Bill Barnes is the chief academic officer for the Howard County Public School System in Maryland. He is also the second vice president of the NCSM and has served as an adjunct professor for Johns Hopkins University, the University of Maryland–Baltimore County, McDaniel College, and Towson University.

Barnes is passionate about ensuring equity and access in mathematics for students, families, and staff. His experiences drive his advocacy efforts as he works to ensure opportunity and access to underserved and underperforming populations. He fosters partnership among schools, families, and community resources in an effort to eliminate traditional educational barriers.

A past president of the Maryland Council of Teachers of Mathematics, Barnes has served as the affiliate service committee eastern region 2 representative for the NCTM and regional team leader for the NCSM.

Barnes is the recipient of the 2003 Maryland Presidential Award for Excellence in Mathematics and Science Teaching. He was named Outstanding Middle School Math Teacher by the Maryland Council of Teachers of Mathematics and Maryland Public Television and Master Teacher of the Year by the National Teacher Training Institute.

Barnes earned a bachelor of science degree in mathematics from Towson University and a master of science degree in mathematics and science education from Johns Hopkins University.

To learn more about Bill Barnes's work, follow him on Twitter @BillJBarnes.

About the Authors

Sarah Schuhl is an educational coach and consultant specializing in mathematics, professional learning communities, common formative and summative assessments, school improvement, and response to intervention (RTI). She has worked in schools as a secondary mathematics teacher, high school instructional coach, and K–12 mathematics specialist.

Schuhl was instrumental in the creation of a PLC in the Centennial School District in Oregon, helping teachers make large gains in student achievement. She earned the Centennial School District Triple C Award in 2012.

Sarah designs meaningful professional development in districts throughout the United States focused on strengthening the teaching and learning of mathematics, having teachers learn from one another when working effectively as a collaborative team in a PLC, and striving to ensure the learning of each and every student through assessment practices and intervention. Her practical approach includes working with teachers and administrators to implement assessments for learning, analyze data, collectively respond to student learning, and map standards.

Since 2015, Schuhl has coauthored the books *Engage in the Mathematical Practices: Strategies to Build Numeracy and Literacy With K–5 Learners* and *School Improvement for All: A How-to Guide for Doing the Right Work.*

Previously, Schuhl served as a member and chair of the National Council of Teachers of Mathematics (NCTM) editorial panel for the journal *Mathematics Teacher*. Her work with the Oregon Department of Education includes designing mathematics assessment items, test specifications and blueprints, and rubrics for achievement-level descriptors. She has also contributed as a writer to a middle school mathematics series and an elementary mathematics intervention program.

Schuhl earned a bachelor of science in mathematics from Eastern Oregon University and a master of science in mathematics education from Portland State University.

To learn more about Sarah Schuhl's work, follow her on Twitter @SSchuhl.

Mona Toncheff, an educational consultant and author, is also currently working as a project manager for the Arizona Mathematics Partnership (a National Science Foundation–funded grant). A passionate educator working with diverse populations in a Title I district, she previously worked as both a mathematics teacher and a mathematics content specialist for the Phoenix Union High School District in Arizona. In the latter role, she provided professional development to high school teachers and administrators related to quality mathematics teaching and learning and working in effective collaborative teams.

As a writer and consultant, Mona works with educators and leaders nationwide to build collaborative teams, empowering them with effective strategies for aligning curriculum, instruction, and assessment to ensure all students receive high-quality mathematics instruction.

Toncheff is currently an active member of the National Council of Supervisors of Mathematics (NCSM) board and has served NCSM in the roles of secretary (2007–2008), director of western region 1 (2012–2015), second vice president (2015–2016), first vice president (2016–2017), marketing and e-news editor (2017–2018), and president-elect (2018–2019). In addition to her work with NCSM, Mona is also the current president of Arizona Mathematics Leaders (2016–2018). She was named 2009 Phoenix Union High School District Teacher of the Year; and in 2014, she received the Copper Apple Award for leadership in mathematics from the Arizona Association of Teachers of Mathematics.

Toncheff earned a bachelor of science degree from Arizona State University and a master of education degree in educational leadership from Northern Arizona University.

To learn more about Mona Toncheff's work, follow her on Twitter @toncheff5.

To book Timothy D. Kanold, Jessica Kanold-McIntyre, Matthew R. Larson, Bill Barnes, Sarah Schuhl, or Mona Toncheff for professional development, contact pd@SolutionTree.com.

Preface

By Timothy D. Kanold

In the early 1990s, I had the honor of working with Rick DuFour at Adlai E. Stevenson High School in Lincolnshire, Illinois. During that time, Rick—then principal of Stevenson—began his revolutionary work as one of the architects of the Professional Learning Communities at Work® (PLC) process. My role at Stevenson was to initiate and incorporate the elements of the PLC process into the K–12 mathematics programs, including the K–5 and 6–8 schools feeding into the Stevenson district.

In those early days of our PLC work, we understood that grade-level or course-based mathematics collaborative teacher teams provided us a chance to share and become more transparent with one another. We exchanged knowledge and reflected on our growth and improvement as teachers in order to create and enhance student agency for learning mathematics. As colleagues and team members, we taught, coached, and learned from one another; we refined our mathematics teaching practice and took action to improve. However, we had one major secret we kept from one another. We did not know our mathematics *instruction* story. We did not have much clarity on an instruction vision that would improve the learning of our students in mathematics and cause greater student agency.

We did not initially understand how the work of our collaborative teacher teams—especially in mathematics at all grade levels—when focused on the right instruction criteria, could erase inequities in student learning that the wide variance of our professional assessment practice caused.

Through our work together, we realized that, without intending to, we often were creating massive gaps in student learning because of our isolation from one another; our isolated decisions about the specifics of lesson design and the teaching of mathematics were a crushing consequence in a vertically connected curriculum like mathematics.

We also could not have anticipated one of the best benefits of working in community with one another: the benefit of belonging to something larger than ourselves. There is a benefit to learning about various teaching and assessing strategies from each other, *as professionals*. We realized it was often in community we found deeper meaning to our work, and strength in the journey as we solved the complex issues we faced each day and each week of the school season, *together*.

The idea of this collaborative focus to the real work we do as mathematics teachers is at the heart of the *Every Student Can Learn Mathematics* series. The belief, that if we do the right work together, then just maybe every student can be inspired to learn mathematics, has been the driving force of our work for more than thirty years. And thus, the title of this mathematics professional development series was born.

In this series, we emphasize the concept of *team action*. We recognize that some readers may be the only members of a grade level or mathematics course. In that case, we recommend you work with a colleague a grade level or course above or below your own. Or, work with other job-alike teachers across a geographical region as technology allows. Collaborative teams are the engines that drive the PLC process.

A PLC in its truest form is "an ongoing process in which educators work collaboratively in recurring cycles of collective inquiry and action research to achieve better results for the students they serve" (DuFour, DuFour, Eaker, Many, & Mattos, 2016, p. 10). This book and the other three in the *Every Student Can Learn Mathematics* series feature a wide range of voices, tools, and discussion protocols offering advice, tips, and knowledge for your PLC-based collaborative mathematics team.

The coauthors of the *Every Student Can Learn Mathematics* series—Bill Barnes, Matt Larson, Jessica Kanold-McIntyre, Sarah Schuhl, and Mona Toncheff—have each been on their own journeys with the deep and collaborative work of PLCs for mathematics. They have all spent significant time in the classroom as highly successful practitioners, leaders, and coaches of K–12 mathematics teams, designing and leading the structures and the culture necessary for effective and collaborative team efforts. They have lived through and led the mathematics professional growth actions this book advocates within diverse K–12 educational settings in rural, urban, and suburban schools.

In this book, we tell our mathematics *instruction* story. It is a story about your daily choices for the balance of higher-level and lower-level-cognitive-demand mathematical tasks to teach the essential learning standards of the unit, your strategies used for teaching those standards, the daily discourse balance in your mathematics class, and the effective ways in which every mathematics lesson should begin and end. It is a K–12 story that, when well implemented, will bring great satisfaction to your work as a mathematics professional and result in a positive, persevering, and formative learning impact on your students.

For your grade-level or course-based collaborative team, helping students to persevere during a mathematics lesson as *part of a formative process* for learning is where the most impactful work of your actual teaching and student learning and meaning is located. This formative process is at the heart of *Mathematics Instruction and Tasks in a PLC at Work*. We want to help your team participate in meaningful discussions about how to prepare and then execute a mathematics lesson design that will significantly increase student learning and effort each day.

We hope you will join us in the journey of significantly improving student learning in mathematics by leading and improving your mathematics instruction story for your team, your school, or your district. The conditions and the actions for adult learning of mathematics *together* reside in the pages of this book. We hope the stories we tell, the tools we provide, and the opportunities for reflection serve you well in your daily professional work in a discipline we all love—mathematics!

Introduction

If you are a teacher of mathematics, then this book is for you! Whether you are a novice or a master teacher; an elementary, middle, or high school teacher; a rural, suburban, or urban teacher; this book is for you. It is for all teachers and support professionals who are part of the K–12 mathematics learning experience.

Teaching mathematics so *each and every student* learns the K–12 college-preparatory mathematics curriculum, develops a positive mathematics identity, and becomes empowered by mathematics is a complex and challenging task. Trying to solve that task in isolation from your colleagues will not result in erasing inequities that exist in your schools. The pursuit and hope of developing into a collaborative community with your colleagues and moving away from isolated professional practice are necessary, hard, exhausting, and sometimes overwhelming.

Your professional life as a mathematics teacher is not easy. In this book, you and your colleagues will focus your time and energy on collaborative efforts that result in significant improvement in student learning, as students participate in the formative learning process of *reflect, refine, and act* over and over again throughout the school year.

Some educators may ask, "Why become engaged in collaborative mathematics teaching actions in your school or department?" The answer is simple: *equity*.

What is equity? To answer that, it is helpful to first examine inequity. In traditional schools in which teachers work in isolation, there is often a wide discrepancy in teacher practice. Teachers in the same grade level or course may teach, assess, assign homework, and grade students in mathematics quite differently—there may be a lack of rigor consistency in what teachers expect students to know and be able to do, how they will know when students have learned, what they will do when students have not learned, and how they will proceed when students have demonstrated learning. Such wide variance in potential teacher practice among grade-level and course-based teachers then causes inequities as students pass from course to course and grade to grade.

These types of equity issues require you and your colleagues to engage in team discussions around the development and use of assessments that provide evidence of and strategies for improving student learning.

Equity and PLCs

The PLC at Work® process is one of the best and most promising models your school or district can use to build a more equitable response for student learning. The architects of the PLC process, Richard DuFour, Robert Eaker, and Rebecca DuFour, designed the process around three big ideas and four critical questions that placed learning, collaboration, and results at the forefront of our work (DuFour, et al., 2016). As DuFour, Eaker, and DuFour explain in their large cadre of work, schools and districts that commit to the PLC transformation process rally around the following three big ideas (DuFour et al., 2016).

1. **A focus on learning:** Teachers focus on learning as the fundamental purpose of the school rather than on teaching as the fundamental purpose.
2. **A collaborative culture:** Teachers work together in teams interdependently to achieve a common goal or goals for which members are mutually accountable.
3. **A results orientation:** Team members are constantly seeking evidence of the results they desire—high levels of student learning.

Additionally, teacher teams within a PLC focus on four critical questions (DuFour et al., 2016) as part of their instruction and task-creation routines used to inspire student learning:

1. What do we want all students to know and be able to do *in class*?
2. How will we know if they learn it *in class*?
3. How will we respond *in class* when some students do not learn?
4. How will we extend the learning *in class* for students who are already proficient?

The four critical PLC questions provide an equity lens for your professional work during instruction. Notice the intentional adaption of the four critical questions around the words *in class*. This is intentional, as this book is all about the student learning process in class during the lesson and the potential gaps that will exist if you and your colleagues do not agree on the rigor for the mathematical tasks you use to answer the question, What do we want all students to know and be able to do in class today?

Imagine the devastating effects on students if you do not reach team agreement on the lesson-design criteria and routines used during the lesson (see critical question 2) as you engage your students in the mathematics lesson each day. Imagine the lack of student agency (their voice, ownership, perseverance, and action during learning) if you do not work together to create a unified, robust formative process for helping students *own* their response during class when they are and are not learning (PLC critical questions three and four).

For you and your colleagues to answer these four PLC critical questions well during the lesson requires the development, use, and understanding of lesson-design criteria that will cause students to engage in the lesson, persevere through the lesson, and embrace their errors as they demonstrate learning pathways for the various mathematics tasks you present to them.

The concept of your team *reflecting together and then taking action* around the right mathematics lesson-design work is an emphasis in the *Every Student Can Learn Mathematics* series. The potential actions you and your colleagues take together improve the likelihood of more equitable mathematics learning experiences for every K–12 student.

The Reflect, Refine, and Act Cycle

Figure I.1 illustrates the reflect, refine, and act cycle, our perspective about the process of lifelong learning—for you, and for your students. The very nature of the profession is about the development of skills toward learning. Those skills are part of an ongoing process you pursue together with your colleagues.

More important, the reflect, refine, and act cycle is a *formative* learning cycle described throughout all four books in the series. When you embrace mathematics learning as a *process*, you and your students:

- **Reflect**—Work the task, and then ask: "Is this the best solution strategy?"

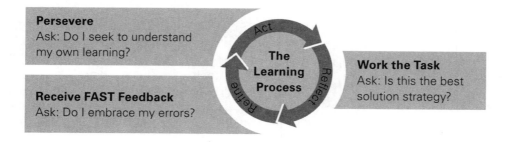

Figure I.1: Reflect, refine, and act cycle for formative student learning.

- **Refine**—Receive FAST feedback and ask, "Do I embrace my errors?"
- **Act**—Persevere and ask, "Do I seek to understand my own learning?"

The intent of this *Every Child Can Learn Mathematics* series is to provide you with a systemic way to structure and facilitate deep team discussions necessary to lead an effective and ongoing adult and student learning process each and every school year.

Team Actions and the Mathematics in a PLC at Work Framework

The *Every Student Can Learn Mathematics* series has four books that focus on a total of six teacher team actions and two mathematics coaching actions within four primary categories.

1. *Mathematics Assessment and Intervention in a PLC at Work*
2. *Mathematics Instruction and Tasks in a PLC at Work*
3. *Mathematics Homework and Grading in a PLC at Work*
4. *Mathematics Coaching and Collaboration in a PLC at Work*

Figure I.2 (page 4) shows each of these four categories and the two actions within them. These eight actions focus on the nature of the professional work of your teacher teams and how they should respond to the four critical questions of a PLC (DuFour et al., 2016).

So, who exactly should be working with you on a collaborative team to develop high-quality, essential, and balanced lesson-design elements and then use the lesson-design elements to provide formative feedback and student perseverance? With whom does it make the most sense for you to collaborate and learn to fulfill team actions 3 and 4 from figure I.2?

Most commonly, a collaborative team consists of two or more teachers who teach the same grade level or course. Through your focused work addressing the four critical questions of a PLC, you provide every student in your grade level or course with equitable learning experiences and expectations, opportunities for sustained perseverance, and robust formative feedback during the lesson, regardless of the teacher he or she receives.

If, however, you are a singleton (a lone teacher who does not have a colleague who teaches the same grade level or course), you will have to determine who it makes the most sense for you to work with as you strengthen your lesson design and student feedback skills. Leadership consultant and author Aaron Hansen (2015) suggests the following possibilities for creating teams for singletons.

- Vertical teams (for example, a primary school team of grades K–2 teachers or a middle school mathematics department team for grades 6–8)
- Virtual teams (for example, a team comprising teachers from different sites who teach the same grade level or course and collaborate virtually with one another across geographical regions)
- Grade-level or course-based team *expansion* (for example, a team of grade-level or course-based teachers in which each teacher teaches all sections of grade 6, grade 7, or grade 8; the teachers expand to teach and share two or three grade levels instead of only one in order to create a grade-level or course-based team)

About This Book

Every grade-level or course-based collaborative team of mathematics teachers in a PLC culture is expected to meet on an ongoing basis to discuss how its mathematics lessons are designed to ask and answer the four PLC critical questions as students are learning during class. In this book in the series, you explore two specific team actions for your professional work.

- **Team action 3:** *Develop* high-quality mathematics lessons for daily instruction.
- **Team action 4:** *Use* effective lesson designs to provide formative feedback and student perseverance.

You might be surprised, but there is a theme that runs through mathematics instruction and lesson design when working as part of a collaborative mathematics team within a PLC at Work culture. Ready?

It's *balance and perseverance.*

Every Student Can Learn Mathematics Series Team and Coaching Actions Serving the Four Critical Questions of a PLC at Work	1. What do we want all students to know and be able to do?	2. How will we know if they learn it?	3. How will we respond when some students do not learn?	4. How will we extend the learning for students who are already proficient?
Mathematics Assessment and Intervention in a PLC at Work				
Team action 1: Develop high-quality common assessments for the agreed-on essential learning standards.	■	■		
Team action 2: Use common assessments for formative student learning and intervention.			■	■
Mathematics Instruction and Tasks in a PLC at Work				
Team action 3: Develop high-quality mathematics lessons for daily instruction.	■	■		
Team action 4: Use effective lesson designs to provide formative feedback and student perseverance.			■	■
Mathematics Homework and Grading in a PLC at Work				
Team action 5: Develop and use high-quality common independent practice assignments for formative student learning.	■	■		
Team action 6: Develop and use high-quality common grading components and formative grading routines.			■	■
Mathematics Coaching and Collaboration in a PLC at Work				
Coaching action 1: Develop PLC structures for effective teacher team engagement, transparency, and action.	■	■		
Coaching action 2: Use common assessments and lesson-design elements for teacher team reflection, data analysis, and subsequent action.			■	■

Figure I.2: Mathematics in a PLC at Work framework.

*Visit **go.SolutionTree.com/MathematicsatWork** for a free reproducible version of this figure.*

Your daily lesson design and planning for a mathematics lesson can easily fall into a routine that is unbalanced in its mathematical task selection, strategies used to teach the lesson, and student discourse and engagement during the lesson. Without ongoing team discussion with your colleagues about your daily lesson design, you can unintentionally cause deep inequities in student learning.

Do you know the following daily lesson routines of your colleagues?

- Do you each declare the mathematics standard to be learned each day?
- Do you each connect every mathematics task used during the lesson to the standard for the day?
- Do you each balance the use of lower-level-cognitive-demand tasks (procedural knowledge with rote routines) with the use of higher-level, open-ended mathematical tasks?

- Do you each use application and mathematical modeling tasks during the unit?
- Do you each teach the academic vocabulary of the daily lesson?
- Do you each use a formative learning process that actively engages students during the lesson?
- Do you each use technology or other mathematical models as a routine part of the lesson design?

Wide variances in your daily decision making can cause a *rigor* inequity for students in the same grade level or course. In a vertically connected curriculum like mathematics this variance can cause learning gaps as students progress through the grades.

Significant lesson-planning differences may exist with how the lesson begins and ends, as well. You may use prior-knowledge warm-up activities every day with a student-led closure activity. However, your colleagues may not.

Mathematics lessons then have a lot of daily choices you must make. And those choices should be designed to help your students demonstrate "productive perseverance" during a mathematics lesson and persevere through the variety of mathematics tasks they must do to demonstrate their learning (M. Larson, personal communication, July 30, 2017).

In this book, *Mathematics Instruction and Tasks in a PLC at Work*, there is intentional guidance to help you and your colleagues reflect on your current lesson-design elements, compare your current practice against high-quality standards of mathematics lesson design, and then develop and use lessons that effectively engage students with those lesson elements.

The benefit of these lesson-design elements will be improved student perseverance in class, and they are most likely to result in retention of learning the expected mathematics standards for your grade level or course.

In this book, you will find spaces to write out reflections about your practice. You are also provided team discussion protocol tools to make your team meeting discussions focused, mindful, and meaningful.

The team discussion tools and protocols are designed for you to eventually feel confident and comfortable in conversations with one another about your lesson content and process, and in moving toward greater transparency in your instructional practice and understanding of the standards with colleagues. In this book, you will also find personal stories from the authors' experiences that shed light on the impact of your team actions on classroom practice.

This book is divided into two parts. Part 1 focuses on the third team action—*Develop* high-quality, essential, and balanced lesson-design elements. The chapters in part 1 explore six research-affirmed lesson-design elements for highly effective daily mathematics lessons. The final chapter in part 1 presents the Mathematics in a PLC at Work lesson-design tool that helps ensure your team reaches *daily and unit* mathematics lesson clarity on all four of the PLC critical questions. Part 2 focuses on the fourth team action—*Use* the lesson-design elements to provide formative feedback and sustained student perseverance during the lesson. The chapters in part 2 explore the *how* of the lesson-design process using the six essential lesson-design elements.

This *Every Student Can Learn Mathematics* professional development series is steeped in the belief that as classroom teachers of mathematics, your decisions and your daily actions matter. You have the power to decide and choose the mathematical tasks students will be required to perform during the lesson, during the homework you develop and design, during the unit assessments such as quizzes and tests you design, and during projects and other high-performance tasks. You have the power to decide the nature of the rigor for those mathematical tasks, the nature of the student communication and discourse to learn those tasks, and the nature of whether or not learning mathematics should be a formative feedback process for you and your students.

You can visit **go.SolutionTree.com/Mathematicsat Work** to access the free reproducibles listed in this book. In addition, online you will find grade-level lesson-design samples along with a comprehensive list of free online resources—"Online Resources Reference Guide for Mathematics Support"—to support your work in mathematics teaching and learning.

Most important, you have the power to decide if you will do all of this challenging mathematics work of your profession alone or with others. As you embrace the belief that together the work of your PLC can overcome the many obstacles you face each day, then *every student can learn mathematics* just may become a reality in your school.

PART 1

Team Action 3: Develop High-Quality Mathematics Lessons for Daily Instruction

The function of education is to teach one to think intensively and to think critically.

—Martin Luther King Jr.

You and your teaching colleagues have a lot of power—power that resides in the lesson-design decisions you make each and every day. In some ways, Martin Luther King Jr. provides a description central to the essence of great mathematics instruction with his powerful quote. Admittedly, in the 21st century education environment, teachers would use the word *persevere* instead of *think intensively* and the phrase *higher-level-cognitive-demand reasoning* instead of *think critically*. And in every lesson design and instructional moment, your students should do these actions with *balance*. As you and your colleagues prepare mathematics lessons each day, and for each unit of study, your lesson-design choices shape—and can make or break—students' mathematical learning experiences.

In part 1 of this book, you focus on developing high-quality and balanced lesson-design elements—team action 3. You and your team explore the most essential elements of lesson-planning design, including the nature of the essential learning standards and the mathematical tasks you choose to teach each daily standard. This action is about preparing each of six lesson-design elements in advance of the lesson to significantly improve student learning. In some cases, effective lesson design, coupled with effective lesson implementation, can double the speed of learning toward the essential mathematics standards of the unit (Popham, 2011).

You will also be asked to reflect on and refine your lesson routines in order to achieve such a lofty expectation. You will reflect on how your current lessons become more coherent by meeting the demands of the four critical questions of a PLC (DuFour et al., 2016) discussed in the introduction to this book.

Your Mathematics Instruction *Purpose*

To begin, think about how you approach your mathematics lesson preparation each day. Regardless of the nature of the mathematics lesson you are about to prepare, it is helpful to explore the phrases *relevant mathematics curriculum* and *meaningful mathematics curriculum*. Are these the same things? Or are they different? Why do these terms matter when thinking about your lesson design?

> **TEACHER** *Reflection*
>
> Take a moment to reflect on and discuss with a colleague the difference between a *relevant* mathematics curriculum and a *meaningful* mathematics curriculum.
>
> _____
> _____
> _____
> _____
> _____
> _____
> _____
> _____

Gaining an understanding of the K–12 mathematics curriculum for 2020 and beyond requires an understanding of the difference between what is meant by a *relevant* mathematics curriculum and what is meant by a *meaningful* mathematics curriculum. These phrases in turn force you to analyze and review the elements of mathematics lesson design that reflect a balance of learning *tasks and discourse* for the students, and elements that cause your students to *persevere and embrace errors* made when practicing during the lesson.

Although some educators use the words *relevant* and *meaningful* interchangeably, these words support significantly different ideals for mathematics lesson design.

Relevant Mathematics

Think of a relevant mathematics curriculum as one that contains "the important stuff"—that which establishes a context for the learning of the lesson. *Relevant* mathematics is mathematics units of study that contain essential mathematics and mathematical tasks students need to know. To determine if a lesson is relevant, ask, "Does the lesson present important and essential mathematics? *Why* must students learn this mathematics lesson, today?"

Think of relevant mathematics as establishing a *context* for learning the mathematics standard of the daily lesson. Meaningful mathematics, on the other hand, presents a different type of lesson-design challenge.

Meaningful Mathematics

Think of meaningful mathematics as an aspect of your lesson design that is all about the student's point of view and engagement during the lesson. *Meaningful* mathematics is mathematics that contains elements that create student agency in learning through reasoning and sense-making, while also connecting to the students' prior knowledge and understanding. To determine if a lesson is meaningful, ask, "How does the daily lesson design create meaning for the student? What will students be *doing* during each part of the lesson? And, will students persevere when they get stuck?"

In addition, meaningful mathematics is grounded in your daily choice of mathematical tasks and ways in which the tasks draw on multiple sources of knowledge from the student. This includes intentionally tapping into students' prior knowledge and experiences, including their cultural, linguistic, family, and community resources as you teach. This is one of the five equity-based teaching practices Julia Aguirre, Karen Mayfield-Ingram, and Danny Bernard Martin (2013) offer. When teachers draw on these resources to make mathematics meaningful for students, they help "students bridge everyday experiences to learn mathematics" (Aguirre et al., 2013, p. 43).

As you understand that tasks form the basis for students' opportunities to learn what mathematics is and how *one does* mathematics, it creates a realization that your task selection for the lesson is of high priority and importance. The level and kind of thinking required by the mathematical instructional tasks you choose ultimately influence what your students learn that day.

In the sections that follow, you will have the opportunity to reflect on your current lesson planning and design practice. Then you will examine six essential elements of every mathematics lesson you design and ultimately use with your students each day.

The Mathematics in a PLC at Work Instructional Framework

Think about the lessons you taught during your most recent unit of mathematics for your grade level or course. How did you design your lesson to ensure it was both *relevant* and *meaningful* for your students? Think about the most essential standards you were trying to help students learn. Use the seven questions in figure P1.1 to focus your teams' lesson-design discussions.

The real purpose of any mathematics lesson is about facilitating your students' learning of the essential standard or learning target for that day. The choices you make regarding the mathematics problems and tasks are the means for how you get your students to the final destination: *learning the essential standards for the unit*. How you will use those mathematical task choices in class becomes one of your greatest professional responsibilities.

Another equity-based mathematics teaching practice offered by Aguirre et al. (2013) is going deep with mathematics. They specifically argue that mathematics "lessons include high cognitive demand tasks that support and strengthen students' development of the strands of mathematical proficiency" (p. 43).

Directions: Use the following prompts to guide discussion about your current lesson design.

Purpose of the lesson:

1. What is the fundamental purpose of a mathematics lesson?

2. How do you inform your students of the relevance—the *why*—for the day? Do you inform them in writing or verbally?

Essential elements of a mathematics lesson:

Respond to each question with a yes or no, and then briefly explain how you make the lesson-design choice or why you do not make the lesson-design choice.

3. Do you choose a warm-up or prior-knowledge mathematics question or task to begin each lesson?

4. Do you choose to discuss and connect key vocabulary words for the lesson?

5. Do you choose lower-level- and higher-level-cognitive-demand mathematical tasks that align to the essential standard for each lesson? If so, is it a collaborative teacher team activity?

6. Do you intentionally choose whole-group *and* small-group discourse activities as part of the lesson experience?

7. Do you close each lesson with a student-led summary?

Figure P1.1: Team discussion tool—Collaborative lesson-design elements.

Visit go.SolutionTree.com/MathematicsatWork for a free reproducible version of this figure.

Thus a mathematics lesson is about so much more than just "doing" a bunch of mathematics problems or tasks with your students. The choices of the tasks you use to teach the lesson are very important. In fact, they give you a lot of power as a teacher, but they are in and of themselves not the reason for or the purpose of the mathematics lesson.

The true purpose of any mathematics lesson is to *maximize student engagement, communication, and perseverance during the lesson based on the tasks you have chosen*, in order to help students learn the essential learning standards for each unit of mathematics.

The chapters that follow in part 1 provide six research-affirmed lesson-design elements and an instructional framework for highly effective daily mathematics lessons every day. Some of these elements may already be present in your daily planning; you just need to work with your team members to brainstorm and share creative ideas about how the elements work well for you *and* how they can work well for your colleagues too.

Other elements may not yet be present in your lessons. You can decide how to begin to adjust your daily lesson design to better impact student perseverance and learning in your mathematics classroom.

As you and your collaborative team reflect on the elements present in your current lesson design, you can compare your progress on six mathematics lesson-design elements. The following six elements of effective mathematics lesson design are the focus of each of the six chapters in part 1.

1. Essential learning standards: the *why* of the lesson
2. Prior-knowledge warm-up activities
3. Academic language vocabulary as part of instruction
4. Lower- and higher-level-cognitive-demand mathematical task balance
5. Whole-group discourse and small-group discourse balance
6. Lesson closure for evidence of learning

You and your team can use figure P1.2, the lesson-design evaluation tool, to evaluate the quality of your current mathematics lessons. It will help you identify areas of strength and areas for lesson-design growth.

As you begin your team journey through the chapters in part 1, continue to use this lesson-design evaluation tool to assess the strengths and weaknesses of your current instructional planning for mathematics.

Your daily lessons should provide an opportunity for your students to *reflect*, *refine*, and *act* during the lesson. You should expect your students to use the mathematical tasks you have chosen and the feedback you provide during the lesson to refine their errors in the process of learning each day.

This book includes several additional team discussion tools designed to support your team's work in part 1.

- Collaborative Lesson-Design Elements (figure P1.1, page 9)
- Categories of Vocabulary Challenges for Students (figure 3.1, page 29)
- Choosing Mathematical Tasks for Lesson Design During the Unit (figure 4.2, page 37)
- Student Discourse During the Lesson (figure 5.1, page 44)
- Lesson-Closure Reflection (figure 6.1, page 48)
- Protocol for Team Analysis of the Mathematics in a PLC at Work Lesson-Design Tool (figure 7.2, page 55)

As the mathematics lesson unfolds, it is your responsibility to make sure your students move through the four critical questions of a PLC: What is it you want your students to know and be able to do (the essential learning standard for the day)? How will you know if they know it (based on ongoing student performance of the tasks you have chosen during the lesson)? What will be your response if they do or don't know it (formatively scaffolding or advancing student learning during the lessons' tasks)?

Your daily lesson-design preparation throughout the school year has a certain rhythm to it. And that rhythm begins with understanding the essential learning standard, why it is relevant, and how the standard serves the general progression of the mathematics content.

High-Quality Lesson-Design Indicators	Description of Level 1	Requirements of the Indicator Are Not Present	Limited Requirements of This Indicator Are Present	Substantially Meets the Requirements of the Indicator	Fully Achieves the Requirements of the Indicator	Description of Level 4
1. Essential Learning Standards: The Why of the Lesson	The lesson references an essential learning standard but doesn't have a clear learning target, and there is no evidence of consistent standard or target language across the collaborative team.	1	2	3	4	The lesson design declares a daily learning target aligned to an essential learning standard for the unit. Teachers share a context for that learning target with students during the lesson.
2. Prior-Knowledge Warm-Up Activities	There is either no warm-up activity to the lesson content or the warm-up activity exists, but does not clearly support students' accessing prior knowledge needed for the lesson.	1	2	3	4	There is a prior-knowledge warm-up task that includes an opportunity for students to work together and engage in thinking about the mathematics necessary to persevere during the lesson.
3. Academic Language Vocabulary as Part of Instruction	The lesson does not address academic language explicitly with a formal plan for ensuring student clarity.	1	2	3	4	There is evidence of focused vocabulary instruction to support the learning of the mathematics content across grade-level or course-based teams.
4. Lower- and Higher-Level-Cognitive-Demand Mathematical Task Balance	There is no evidence of a balance of lower- and higher-level-cognitive-demand tasks. There are no specific strategies for engaging students in the sense-making or application of the content.	1	2	3	4	There is a balance of higher-level and lower-level-cognitive-demand mathematical tasks within the lesson plan with specific focus on formative routines, feedback from peers, and the teacher during the lesson.
5. Whole-Group and Small-Group Discourse Balance	There are no specific strategies for how students will discuss and share their thinking with their peers. The lesson plan relies solely on whole-group discourse from the front of the classroom with only the teacher evaluating the responses to each student question.	1	2	3	4	There are intentional plans for the type of discourse (whole group or small group) that students will experience for each mathematical task and portion of the lesson. There is a commitment to balancing student time to process and communicate with one another (what you see and hear the students doing) against the time given to teacher-directed instruction.
6. Lesson Closure for Evidence of Learning	The lesson plan includes either no summary or a teacher-led summary of the lesson (as opposed to a student-led summary). There is no opportunity for students to evaluate if they meet and understand the learning target for the day.	1	2	3	4	The lesson includes a student-led closure activity to determine if the lesson helped students understand the learning target or essential learning standard for the day.

Figure P1.2: Mathematics in a PLC at Work instructional framework and lesson-design evaluation tool.

Visit go.SolutionTree.com/MathematicsatWork for a free reproducible version of this figure.

CHAPTER 1

Essential Learning Standards: The *Why* of the Lesson

> When teachers refer to the goals of the lesson during instruction, students become more focused and better able to perform self-assessment and monitor their own learning.
>
> —Shirley Clarke, Helen Timperley, and John Hattie

Perhaps one of the most glaring weaknesses of many mathematics lessons is the lack of student facility to state with clarity, understanding, and precision the actual learning standard for the lesson. If you were to ask most students as they move on to the next activity (elementary school) or walk out of class (middle school or high school) the question, "What did you learn in math class today?" they most likely will respond, "I learned how to do some math problems."

For many students, learning mathematics is about doing a bunch of disjointed and seemingly meaningless mathematical tasks, mathematics problems, mathematics questions, or just "math stuff." Most students will not be able to, when called on, explain the reason they did the mathematics tasks their teacher chose for the lesson that day. They are unable to explain the *why* of the lesson or describe the standard for the lesson.

For example, in a fifth-grade classroom, the intended learning standard might be *I can understand the result when dividing a fraction by another fraction less than or greater than 1*. Yet when asked what they learned today in class, at best, students might say, "We did math problems with fractions."

Pushed further, if students are asked, "Why did you learn this today? What was the purpose of the mathematics lesson?", they generally have no clear response to the question. It just was not part of your lesson design. Thus, knowing the relevance of the lesson and understanding and providing clarity to the mathematics standard are the first elements of lesson planning for you and your team.

The personal story on page 14 illustrates why understanding the essential learning standards is the first of the six essential design elements you and your team will tackle—partially because of the immediate effects on student perseverance and effort and partially because of the forced teacher clarity for understanding the standards.

Developing Clarity

To begin, you and your colleagues should develop clarity on the essential standards for each mathematics unit. This requires your teacher team to break down the three or four essential learning standards for each unit into the daily learning targets for your lessons with an eye toward choosing and aligning the tasks for each lesson.

Essential Learning Standards

For the purposes of this book, the essential learning standards represent the three to four content clusters for the mathematics unit. These are the standards that will be part of your students' assessment analysis for grading purposes described in *Mathematics Assessment and Intervention in a PLC at Work* for the *Every Student Can Learn Mathematics* series. You use the essential learning standards to assess student understanding on all unit quizzes and tests. These clustered mathematics standards act as the driver for organizing your unit assessments and providing student intervention and

Personal Story TIMOTHY KANOLD

When I first got to Stevenson, I was struck by our general lack of knowledge about why we were teaching certain standards each day. It was as if the lessons just fell from the sky with no rhyme or reason. As teachers of mathematics we very rarely expected our students to know why the lesson was so essential—and I am not sure we always had great clarity on why, either. We soon decided that every mathematics lesson we taught would include, in writing, the context of the lesson: *What happened before this lesson, and what will happen after this lesson.*

We noticed an immediate increase in student perseverance and effort during the lesson. We made sure to connect each mathematics task we chose for the lesson directly to the essential learning standard of the lesson; we made a commitment to tell our students the *why* every day. If we were not clear about the standard and understanding the *why*, we asked each other or did some fact finding. Without knowing it, we were into progressions long before the idea of mathematics progressions became popular!

---

support that take place during and after a unit of study. This is why identifying the essential learning standards for the mathematics unit is so important.

However, from a daily lesson-planning perspective, it becomes necessary to unpack those essential learning standards into daily learning targets. Your team "unpacks" the standard by breaking it down into *daily learning targets* that reflect the conceptual knowledge, proficiency skills, and applications necessary for a student to demonstrate understanding of each essential learning standard for the unit. In some cases, an essential mathematics standard may carry over from one unit to the next.

Fortunately, the lesson-design elements that make up part 1 of this book provide a sequential way to think about the elements of your lesson design. These elements serve to help you and your team members further unpack each essential learning standard for the mathematics unit as you consider the academic vocabulary, the cognitive demand of the chosen tasks, and the specific purpose for each daily learning target.

Learning Targets

Learning targets, then, are your daily learning objectives for the lesson. Often there may be several learning targets and mathematics lessons to help you achieve an essential learning standard for the unit. This is why establishing mathematics targets to focus learning "sits at the top" of a framework of research-informed instructional strategies because it signifies "that setting goals [outcomes] is the starting point for all instructional decisions" (Smith, Steele, & Raith, 2017, p. 195). These targets can be written in student-friendly language (*I can* statements). Visit **go.SolutionTree.com/MathematicsatWork** to access and download the tool "K–12 Mathematics Standards in Student- and English Learner–Friendly Language" if you need help stating these standards for each unit and for understanding the standards in more student-friendly terms.

Think of the daily learning targets as a subset of the three to five essential learning standards for the unit. To help you compare and contrast the difference between essential learning standards and daily learning targets, figures 1.1–1.3 (pages 15–17) provide a sample illustration of a second-grade base-ten unit, a seventh-grade statistics unit, and a high school algebra 1 linear equations unit. The essential learning standards in the left column are more formal descriptions, similar to how standards may appear in your state (or province) standards document. In the middle column, the essential learning standard is rewritten as it would appear on the mathematics unit assessments or tests in student-friendly *I can* language. The right column highlights an unwrapping of each standard into daily learning targets you might use for your daily lesson design and planning. It uses a combination of verbs and noun phrases from the original formal essential learning standard.

Your team should establish the connection between the essential learning standards for the unit and the mathematics tasks or problems you chose to teach those standards each day. What is the context for the progression of the standards in the unit? Does each daily learning target reveal a clear purpose as part of a bigger picture of learning? And, moreover, how will you explain and communicate this purpose to the students with clarity?

Essential Learning Standards   15

Grade 2 Apply and Extend Base-Ten Understanding Unit: Essential Learning Standards

Formal Unit Standards (States standard language)	Essential Learning Standards for Assessment and Reflection (Uses student-friendly language)	Daily Learning Targets (Explains what students have to know and be able to do; unwrapped standards)
1. Understand that the three digits of a three-digit number represent amounts of hundreds, tens, and ones; for example, 706 equals 7 hundreds, 0 tens, and 6 ones. Understand the following as special cases: • The number 100 can be thought of as a bundle of ten tens—called a hundred. • The numbers 100, 200, 300, 400, 500, 600, 700, 800, and 900 refer to one, two, three, four, five, six, seven, eight, or nine hundreds (and 0 tens and 0 ones).	I can understand the meaning of the hundreds, tens, and ones in a three-digit number.	• Understand the three digits of a number represent amounts of hundreds, tens, and ones. • Know how to make a hundred using groups of ten. • Demonstrate the value or meaning of 100, 200, 300, 400, 500, 600, 700, 800, and 900 using groups of one hundred.
2. Count within 1,000; skip count by fives, tens, and hundreds.	I can count within 1,000.	• Count within 1,000. • Skip count by fives within 1,000. • Skip count by tens within 1,000. • Skip count by hundreds within 1,000.
3. Read and write numbers to 1,000 using base-ten numerals, number names, and expanded form.	I can read and write numbers using base-ten numerals, number names, and expanded form.	• Read and recognize numbers to 1,000s. • Write numbers to 1,000. • Understand the numerals, number names, and expanded form of numbers to 1,000.
4. Compare two three-digit numbers based on meanings of the hundreds, tens, and ones digits, using >, =, and < symbols to record the results of comparisons.	I can compare two numbers, explain my thinking, and write my answer using >, =, or < symbols.	• Recognize and use the >, =, < symbols to compare numbers. • Recognize the value of a three-digit number using place value understanding.

Figure 1.1: Sample essential learning standards for a grade 2 apply and extend base-ten understanding unit.

Grade 7 Statistics Unit: Essential Learning Standards

Formal Unit Standards (States standard language)	Essential Learning Standards for Assessment and Reflection (Uses student-friendly language)	Daily Learning Targets (Explains what students have to know and be able to do; unwrapped standards)
1. Understand that the probability of a chance event is a number between 0 and 1 that expresses the likelihood of the event occurring. Larger numbers indicate greater likelihood. A probability near 0 indicates an unlikely event, a probability around $\frac{1}{2}$ indicates an event that is neither unlikely nor likely, and a probability near 1 indicates a likely event.	I can explain the meaning of a probability for a chance event.	• Develop a general understanding of the likelihood of events occurring by realizing that probabilities fall between 0 and 1. • Use the terms *likely* and *unlikely* to describe the probability fractions represent. • Understand that a probability near 0 indicates an unlikely event, a probability around $\frac{1}{2}$ indicates an event that is neither unlikely nor likely, and a probability near 1 indicates a likely event. • Understand what the sum of the probability of an event happening and not happening is.
2. Approximate the probability of a chance event by collecting data on the chance process that produces it and observing its long-run relative frequency, and predict the approximate relative frequency given the probability.	I can approximate the probability of a chance event using relative frequency from data collected in a chance process.	• Define a chance event and gather data on the chance process. • Demonstrate understanding of a relative frequency and predict the relative frequency given the probability.
3. Develop a probability model and use it to find probabilities of events. Compare probabilities from a model to observed frequencies; if the agreement is not good, explain possible sources of the discrepancy. • Develop a uniform probability model by assigning equal probability to all outcomes, and use the model to determine probabilities of events. • Develop a probability model (which may not be uniform) by observing frequencies in data generated from a chance process.	I can develop, use, and evaluate probability models.	• Assign equal probability to all outcomes to develop a uniform probability model, and use the model to determine probabilities of events. • Observe frequencies in data generated from a chance process to develop a probability model (which may not be uniform).
4. Find probabilities of compound events using organized lists, tables, tree diagrams, and simulation. • Understand that, just as with simple events, the probability of a compound event is the fraction of outcomes in the sample space for which the compound event occurs. • Represent sample spaces for compound events using methods such as organized lists, tables, and tree diagrams. For an event described in everyday language, identify the outcomes that compose the event in the sample space. • Design and use a simulation to generate frequencies for compound events.	I can find probabilities of compound events using organized lists, tables, tree diagrams, and simulation.	• Understand that, just as with simple events, the probability of a compound event is the fraction of outcomes in the sample space for which the compound event occurs. • Develop probability models to find the probability of simple and compound events (organized lists, tables, tree diagrams, and simulations). • Represent sample spaces for compound events. • Design and use a simulation to generate frequencies for compound events.

Figure 1.2: Sample essential learning standards for a grade 7 statistics unit.

Algebra 1 Linear Equations Unit: Essential Learning Standards

Formal Unit Standards (States standard language)	Essential Learning Standards for Assessment and Reflection (Uses student-friendly language)	Daily Learning Targets (Explains what students have to know and be able to do; unwrapped standards)

1. Interpret expressions that represent a quantity in terms of its context.
  • Interpret parts of an expression, such as terms, factors, and coefficients.
  • Interpret complicated expressions by viewing one or more of their parts as a single entity.
	I can interpret expressions within a context.	• Interpret parts of an expression, such as terms, factors, and coefficients in terms of a given context.
		• Give a context and interpret complicated expressions by viewing one or more of their parts as a single entity.

2. Create equations that describe numbers or relationships. Create equations and inequalities in one variable and use them to solve problems. 3. Solve linear equations and inequalities in one variable, including equations with coefficients represented by letters. 4. Understand solving equations as a process of reasoning and explain the reasoning. Explain each step in solving a simple equation as following from the equality of number asserted at the previous step starting from the assumption that the original equation has a solution. 5. Rearrange formulas to highlight a quantity of interest, using the same reason as in solving equations. For example, rearrange Ohm's law $v = IR$ to highlight resistance R. 6. Use units as a way to understand problems and to guide the solution of multistep problems; choose and interpret units consistently in formulas; choose and interpret the scale and the origin in graphs and data displays. 7. Define appropriate quantities for the purpose of descriptive modeling.	I can create, solve, and reason with linear equations and use appropriate units in context. I can create, solve, and reason with linear inequalities and use appropriate units in context.	• Create one-variable equations and use them to solve problems. • Create one-variable inequalities and use them to solve problems. • Use units as a way to understand and solve multistep problems. • Choose and interpret units in formulas and graphs. • Define quantities. • Explain the steps used to solve simple equations. • Rearrange formulas to highlight a quantity of interest.

Figure 1.3: Sample essential learning standards for an algebra 1 linear equations unit.

For additional unit examples of essential standards with corresponding daily learning targets, visit **go.Solution Tree.com/MathematicsatWork** to access and download first-grade, fourth-grade, and high-school examples. After you read the sample most appropriate to your grade level, answer the questions in the teacher reflection next.

Reflecting on Practice

When you work with your colleagues to examine the essential learning standards for the unit and then determine the aligned learning targets for your daily lessons, you begin to ensure equitable learning expectations and experiences for the students in classrooms across your team.

Take a few moments to reflect on how you and your colleagues develop understanding of each essential learning standard for the mathematics unit and the impact of those standards on your daily learning targets. Consider how many days of instruction you will need for each essential standard based on your understanding of the breakdown for the learning targets.

Remember, answering PLC critical question 1—What is it we want each student to know and be able to do?—is where each lesson starts and ends. Knowing your *why* for the lesson helps clarify the tasks and activities you choose and how you choose to engage students in the learning process. Knowing your *why* addresses the first critical question of a PLC (DuFour et al., 2016): What do we want all students to know and be able to do?

TEAM RECOMMENDATION

Understand the Essential Learning Standards— The *Why* of the Lesson

- As a grade-level or course-based team, identify the common essential learning standards for the unit and prepare the corresponding essential learning targets for each lesson.

- Use the essential learning standards for the unit to help clarify the overarching unit topics, write them as *I can* statements to aid student understanding, and discuss what students must know and be able to do (the daily learning targets) in each lesson to be considered proficient for the standard.

A primary factor in knowing the *why* of a mathematics lesson generates from knowing the context of the lesson. What are the standards that led into this lesson, and where will this lesson help student learning of mathematics a few weeks from now? In the next chapter, you discover how a prior-knowledge warm-up activity not only helps prepare students for entering into the new knowledge of the mathematics lesson for the day, but also helps to create the context for learning.

TEACHER *Reflection*

How do you and your colleagues create, develop, and communicate your understanding of the essential learning standards for the unit?

How do you break down the essential learning standards for the unit into daily learning targets for your lesson-design purposes?

CHAPTER 2

Prior-Knowledge Warm-Up Activities

> Study findings suggest that learning of new content is supported by activating a conceptually relevant prior-knowledge sequence that helps connect the prior knowledge with the new knowledge.
>
> —Pooja G. Sidney and Martha W. Alibali

Your mathematics lesson should always begin with a connection to prior knowledge. Your connection to each student's prior knowledge as they enter into the lesson for the day significantly influences what students learn in the specific situation presented by the mathematical tasks you ask students to pursue. Mathematical tasks can include activities, examples, or problems that your students are to complete as a whole class, in small groups, or individually.

Why should you ensure every mathematics lesson begins with a warm-up prior-knowledge problem or task? The simple answer: *student perseverance.* When you monitor students' initial response to entry concepts into the lesson, they are more likely to persevere longer as the lesson begins (Hattie & Yates, 2014).

Moreover, since the prior-knowledge task is generally a review, it allows you to gather evidence of overall student readiness to begin the lesson. And, finally, it allows you to establish context (or purpose) to the lesson, and, as previously mentioned, it provides an excellent opportunity to draw on multiple resources of knowledge in order to make instruction more meaningful to students (Aguir, et al., 2013).

You can use the prior-knowledge task to create the much-needed context for the *why* of the lesson. You can explain to students about standards they learned in the past and make connections from today's lesson to that prior knowledge. Your "creating context" discussion can come from the prior-knowledge problem and prepares students for the lesson you are about to begin. These types of questions will help you determine how the lesson's learning target fits within the mathematics learning progression of the unit and indicate you are using mathematics outcomes to focus students' learning (NCTM, 2014).

In this chapter, you will learn how to develop this element of your mathematics lesson design. In part 2, chapter 8 of this book (page 67), you will examine *how* to use this essential element of instruction for an effective teacher response to the evidence of learning you collect.

Choosing Prior-Knowledge Warm-Up Activities

You need to be mindful of time as you choose your prior-knowledge warm-up activities. In mathematics lessons, most warm-up activities that activate prior knowledge are generated either by a mathematical task or a discussion prompt you provide for the students as class begins.

A warm-up activity should not take up a lot of class time; no more than five to ten minutes is appropriate. If the warm-up activity is more of an exploration prior-knowledge task (a mathematical task involving investigation rather than questions to answer), then more time might be necessary to allow students to persevere and engage in learning the content in the lesson for the day. In the following personal story, Sarah Schuhl references a conversation with a team about the amount of time spent on a prior-knowledge warm-up task while also discussing the nature or intended purpose of the task.

Personal Story SARAH SCHUHL

I worked with a third-grade team to collaboratively plan a mathematics lesson related to areas of shapes composed of rectangles. When I asked how the lesson would begin with students, the team indicated it always started every lesson with a worksheet from a program that asked students ten random mathematics questions. We examined the questions. The questions for the area lesson ranged from elapsed time to addition with an algorithm to fraction equivalence to pattern recognition. I asked how much of the one-hour lesson was devoted to this warm-up each day. The answers ranged from fifteen to twenty-five minutes. Several teachers explained that students did not know the concepts, and so they retaught the ten problems before starting the lesson. They further explained that the area lesson would be difficult and they wished they had more time.

After much discussion, the teachers finally agreed to write their own warm-up activity. They wrote one question asking students to find the perimeter and area of a rectangle with only two sides labeled so students would practice area of a rectangle as well as be reminded that they can find missing side lengths when they are not included in the diagram. The teachers wrote a second question giving students the area of a rectangle with its base and asking students for the height, again reinforcing the idea that missing side lengths can be determined. They agreed to give students five minutes to work on the problems in their groups and then have two groups share their solutions before launching the lesson with a shape composed of two rectangles and asking students to find its area in their groups.

After the lesson, the teachers were excited with the evidence of student learning in the lesson. They felt the warm-up was quick and activated the knowledge needed to maximize the time spent practicing areas of complex shapes. While spiraled review is important, they decided they will create the warm-ups to prepare students for the learning of the day and might occasionally include one spiral review question, if needed. The team decided it would also work to include the spiral review questions in their homework assignments.

When you are selecting prior-knowledge tasks to support essential learning standards for a lesson, it can be helpful to reference the prior standards as well. These could be standards from a prior grade (or course) to see the level of depth explored the year before. Alternatively, these could be standards from a prior unit. Either way, the goal is to help students connect to the content of the lesson in order to help them make meaning. This helps to clarify where your instruction needs to start and the prior learning you want to connect with through the warm-up activity.

Figure 2.1, figure 2.2, figure 2.3 (page 22), and figure 2.4 (page 23) provide examples of prior-knowledge warm-up mathematics tasks for grades 2, 5, 7, and an algebra 1 high school course. They are based on a standard from the prior grade level. However, keep in mind that sometimes the prior-knowledge mathematical task is from a previous unit or lesson and may not be from a prior grade level, course, or unit. As you choose your warm-up task, make sure the task reflects knowledge most connected to the essential learning standard for the mathematics unit, and the daily learning target to teach those standards you are about to ask students to engage in learning for the day.

Take some time to closely examine the prior-knowledge task most appropriate for your grade level.

Grade or Course	Grade-Level Standard	Prior-Knowledge Standard From Prior Grade, Course, or Unit
Grade 2	**Formal unit standard:** Use addition and subtraction within 100 to solve one- and two-step word problems involving situations of adding to, taking from, putting together, taking apart, and comparing, with unknowns in all positions. **Essential learning standard:** I can use addition and subtraction within 100 to solve one- and two-step word problems. **Daily learning target:** Students will be able to use addition and subtraction to solve one-step word problems involving adding to and putting together.	**Grade 1:** Use addition and subtraction within 20 to solve word problems involving situations of adding to, taking from, putting together, taking apart, and comparing, with unknowns in all positions.

Prior-knowledge task sequence:
1. Pat has eight red flowers and two yellow flowers in a vase. How many flowers are in the vase altogether? Show how you know.
2. There are twelve flowers in a vase. Five are white, and some are purple. How many are purple? Show how you know.

Explanation of task:
By starting with tasks that include numbers under 20, teachers can learn whether or not students understand one- and two-step problems and the strategies that go along with solving them. These prior-knowledge examples specifically focus on putting together and taking apart. Teachers have the opportunity to see how students show their work, especially for question number two. Do students use addition, subtraction, or do they add on?

Figure 2.1: Grade 2 prior-knowledge mathematics task sample.

Grade or Course	Grade-Level Standard	Prior-Knowledge Standard From Prior Grade, Course, or Unit
Grade 5	**Formal unit standard:** Find whole-number quotients of whole numbers with up to four-digit dividends and two-digit divisors, using strategies based on place value, the properties of operations, and/or the relationship between multiplication and division. Illustrate and explain the calculation by using equations, rectangular arrays, and/or area models. **Essential learning standard:** I can find whole-number quotients of whole numbers using multiple strategies. **Daily learning target:** Students will be able to find whole-number quotients of three-digit dividends and two-digit divisors using multiple strategies and explain the calculation using an area model or rectangular array.	**Grade 4:** Find whole-number quotients and remainders with up to four-digit dividends and one-digit divisors, using strategies based on place value, the properties of operations, and/or the relationship between multiplication and division. Illustrate and explain the calculation by using equations, rectangular arrays, and/or area models.

Figure 2.2: Grade 5 prior-knowledge mathematics task sample.

continued →

Prior-knowledge task sequence:
1. Gabriella makes 7 waffles for breakfast. She has 42 berries to put on top of her waffles. She will put an equal number of berries on each waffle. How many berries will Gabriella put on each waffle? Show your work and write the equation.
2. Now suppose Gabriella has 52 berries. What will happen? How many will she put on each waffle and why?
3. Solve 3,476 ÷ 7. Show your work.

Explanation of task:
For the first day of grade 5 instruction on division, knowing the grade 4 standard is four-digit dividend and one-digit divisor, it would be beneficial to start with a task that the majority of grade 5 students will be able to connect to without getting too caught up in errors. This will still give the teacher good insight into student misconceptions and their overall understanding of division. Before asking a rote division question, it helps for you to know if your students can pick up division in context of a word problem.

Grade or Course	Grade-Level Standard	Prior-Knowledge Standard from Prior Grade, Course, or Unit
Grade 7	**Formal unit standard:** Compute unit rates associated with ratios of fractions, including ratios of lengths, areas, and other quantities measured in like or different units. **Essential learning standard:** I can compute unit rates associated with ratios of fractions. **Daily learning target:** Students will be able to compute unit rates associated with ratios of fractions.	**Grade 6:** Understand the concept of a unit rate $\frac{a}{b}$ associated with a ratio $a : b$ with $b \neq 0$, and use rate language in the context of a ratio relationship.

Prior-knowledge task sequence:
What is a unit rate? Describe or provide an example. Share your example with your shoulder partner and be ready to share with the class.
Note: If it seems that the class is struggling to come to consensus on the definition of a unit rate, it may be helpful to have an example ready to help support students' thinking. It would provide an opportunity to have a discussion within a specific context, which can help students who are struggling. For example, Pizza Joint is running a special on pizzas this month: $125 for 10 pizzas. Write a unit rate to express the price for each pizza.

Explanation of task:
Before diving into the grade 7 standard about computing unit rates, it could be helpful to know what students remember about unit rates from grade 6. The question being asked is challenging because students have to recall the information and create their own example, which requires them to work backward and requires them to create. As students are working, the conversations they will be having with their peers should reveal their understanding of the topic. If students struggle through this warm-up, even with a numeric example, then the teacher knows students need to start at the grade 6 level before they can jump into the expectation of the grade 7 standard. If students demonstrate an understanding of unit rates, then the teacher knows to start with more challenging tasks for the lesson.

Figure 2.3: Grade 7 prior-knowledge mathematics task sample.

Grade or Course	Grade-Level Standard	Prior-Knowledge Standard From Prior Grade, Course, or Unit
Algebra 1	**Formal unit standard:** • Understand that the graph of an equation in two variables is the set of all its solutions plotted in the coordinate plane, often forming a curve (which could be a line). • Graph linear functions expressed symbolically and show key features of the graph, by hand in simple cases and using technology for more complicated cases. **Essential learning standard:** I can graph linear and nonlinear functions. **Daily learning target:** • Students will be able to graph different functions by hand. • Students will be able to distinguish between linear and nonlinear functions.	**Grade 8:** • Interpret the equation $y = mx + b$ as defining a linear function, whose graph is a straight line; give examples of functions that are not linear. • Derive the equation $y = mx$ for a line through the origin and the equation $y = mx + b$ for a line intercepting the vertical axis at b.

Prior-knowledge task sequence:

1. Write the equation for the line graphed below.

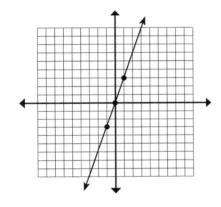

2. Write the equation for the line graphed below.

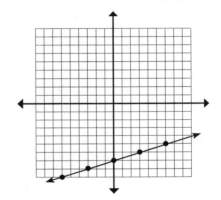

3. What is different about these two graphs? What do you notice about the intercepts and the slopes?

4. Graph an example of a nonlinear function. Be prepared to share with a neighbor.

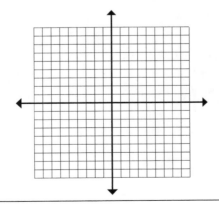

Explanation of task:
The algebra 1 high school standards for this lesson relate to graphing. Given the knowledge students will have based on grade 8 standards, this warm-up starts with the basics of graphing and the vocabulary tied to graphing in order to determine what students remember and can apply. Sample vocabulary might consist of constant rate of change (slope), intercepts, domain, and range. The vocabulary word *function* is new to the high school standards.

Figure 2.4: Algebra 1 high school prior-knowledge mathematics task sample.

In the grade-level examples provided in figures 2.1, 2.2, 2.3, and 2.4, you will notice the use of a sequence of questions to gather evidence of prior knowledge. However, it is also possible that the warm-up is an exploration-type question that students work on together and investigate, based on prior knowledge, without having to answer a series of questions. These types of warm-up problems can help students establish a *why* for the day's learning targets as well.

For example, in a grade 3 lesson introducing division, you might start with some multiplication problems that ask students to think about the number of groups and the number of objects within the group. You could ask students to create a word problem context for the product of 4×6 or 9×7. They could draw or write out verbal examples. Then as you introduce the concept of division, you can tie in the same language used in the warm-up activity to help students make the connection between division and multiplication.

In an algebra 1 course, as the warm-up mathematical task you might ask students to solve a system of equations written in standard form, knowing the only strategy they currently can use for solving a system is graphing. This type of task might seem long and tedious to the students, but it builds an efficiency *reason* to learn and identify a more effective strategy like substitution or elimination. It starts to build the representation context for why you would choose between graphing, substitution, or elimination methods as a solution pathway.

Reflecting on Practice

Now, take a few moments to consider the teacher reflection for the warm-up activities you choose for your lessons and how you determine which tasks and problems to use.

> **TEACHER** *Reflection*
>
> What is the best strategy you currently use to determine the prior-knowledge warm-up activities to use before your lessons? Do you prefer to use mathematical tasks for your warm-ups, discussion prompts, or both?
>
> _____
> _____
> _____
> _____
> _____
> _____
> _____
> _____
> _____

As a team, identify an essential learning standard in your next unit of study. Use figure 2.5, the Prior-Knowledge Task-Planning Tool, to identify either the prior grade or unit essential standard you might need to reference, and choose either a mathematical task or discussion prompt that ties into the prior-knowledge skills that will help students access the essential learning standard for the lesson. In the explanation section, make some notes about why you chose the specific prior-knowledge task or tasks.

At first, it is helpful to write an explanation, but as you progress you may find it is just as beneficial to verbally discuss the reasoning.

Grade or Course	Grade-Level Standard	Prior-Knowledge Standard From Prior Grade, Course, or Unit
	Formal unit standard: Essential learning standard: Daily learning target:	

Prior-knowledge task sequence:

Explanation of task:

Figure 2.5: Prior-knowledge task-planning tool.

*Visit **go.SolutionTree.com/MathematicsatWork** for a free reproducible version of this figure.*

As you and your team consider the prior-knowledge activities, it's important to remember that the mathematical tasks you choose and the discussion prompts you provide need to connect to the conceptual understanding you are trying to build. This is why it is so important to reference prior standards that align to the same strand in the same grade or from a previous grade. This helps you and your students understand and observe the vertical connections and progressions between the various units of content across grades and courses.

Thus, lesson design begins by identifying the essential learning standards for the unit and how the learning target for each day connects to that standard. The next step is identifying a prior-knowledge warm-up task connecting the mathematical task to the *why* of the lesson. These first two essential lesson-design elements establish the relevance for learning the mathematics of the day and address the first critical question of a PLC (DuFour et al., 2016): What do we want all students to know and be able to do?

So what is next? As the mathematics lesson unfolds, you cannot forget about the importance of the academic language vocabulary you will incorporate into your lesson and unit planning and the potential barriers to learning students can face if you fail to address vocabulary as part of the lesson design.

TEAM RECOMMENDATION

Use Prior-Knowledge Warm-Up Activities

- When planning prior-knowledge warm-up activities, be sure to choose mathematical tasks and discussion prompts that connect to the conceptual understanding you are trying to develop for the essential learning standard and learning target that day.

- Use prior-knowledge activities to help connect to the *why* of the lesson, paint a picture of where students are headed in the lesson, and develop student perseverance during the lesson (by reminding them throughout the lesson how chosen activities connect to the learning standard).

CHAPTER 3

Academic Language Vocabulary as Part of Instruction

Students need to know the meaning of mathematics vocabulary words—whether written or spoken—in order to understand and communicate mathematical ideas.

—Rheta N. Rubenstein and Denisse R. Thompson

Academic language is the language students use to make sense of the mathematics they are learning. The language demands of a mathematical task include listening and reading the task to make sense of what the task is asking a student to *do*. In mathematics, academic language vocabulary includes the meaning of words, symbols, notations, and abbreviations within the context of the essential learning standards and potentially linked to prior knowledge.

Academic language vocabulary is important to students' ability to engage verbally and in writing to communicate with their peers to evaluate their progress on mathematical tasks, analyze their own thinking, construct an argument, or justify their reasoning. Despite the importance of mathematics vocabulary, it is an often-overlooked aspect of the lesson-design process.

This is understandable to some extent, as the lesson tends to be about the mathematics the student needs to do in order to learn the standard. And yet words like *denominator, contradict, area, mean ratio, equivalence, tangent,* and *slope field* can confuse many students when it comes to the intent of the lesson.

In order for students to communicate about their ideas orally and in writing, they need a strong mathematics vocabulary. According to Meir Ben-Hur (2006), Senior International Leader for Feuerstein Institute, "Students who lack the formal language of mathematics have difficulties reasoning and communicating about mathematics" (p. 67). Emerging English learners also need to engage in mathematical discourse. Students learning English need opportunities to learn mathematics vocabulary, engage in discourse, make conjectures, and explain their thinking (Civil & Turner, 2014).

At your next team meeting, ask your team members to each share how they currently support developing academic language and vocabulary instruction *during* a lesson. You can use the teacher reflection as a prompt for the discussion.

> **TEACHER** *Reflection*
>
> What strategies and tools do you currently utilize on a consistent basis to ensure each student understands the lesson's vocabulary?
>
> _____
> _____
> _____
> _____
> _____
> _____
> _____
> _____

To help your students learn the content tied to the essential learning standards for the unit, you intentionally plan time for language and vocabulary instruction to support the overarching goal of your instructional plan. Mathematics is communicated by means of a powerful language whose vocabulary must be learned. The ability to reason about and justify mathematical statements is fundamental, as is the ability to use terms and notation with appropriate degrees of precision (Ball et al., 2005).

Incorporating Academic Language Vocabulary

Joan Kenney and her colleagues (2005) indicate that mathematics textbooks expect students to use a vocabulary for understanding both words and symbols:

> Research has shown that mathematics texts contain more concepts per sentence and paragraph than any other type of text. They are written in a very compact style; each sentence contains a lot of information, with little redundancy. The text can contain words as well as numeric and non-numeric symbols to decode. . . . There may also be graphics that must be understood for the text to make sense. (Kenney, Hancewicz, Heuer, Metsisto, & Tuttle, 2005, p. 11)

Students not only need to learn the words but also the symbols that go along with the mathematics. Providing definitions during instruction is not enough to support true understanding of the meaning of the mathematical vocabulary.

Mary Lee Barton and Clare Heidema (2000) further indicate:

> Reading math means decoding and comprehending not only words but also mathematical signs and symbols. Students have to switch between skills used to decode words and skills used to decode symbols. Students have to learn the meaning of each symbol and connect the conceptual idea the symbol represents with the words. (p. 8)

The mathematical vocabulary students encounter can cause confusion for a variety of reasons, challenging student access to the content. For example, think about the word *meter*. In a lesson or unit on measurement, you start students with the root word and then continue on to other words such as *millimeter*, *centimeter*, *decimeter*, and *kilometer*. However, then in the geometry unit, you teach the *perimeter*. Students might ask, "How many in a peri?" This is an example of where the vocabulary and language of mathematics can get confusing for some students, especially if your lesson design is not intentional about supporting student meaning making for their own organization of ideas. Intentionally teaching students the origin of the prefix *peri* can support mathematical meaning later in the lesson.

One way to help organize different types of vocabulary words is to think about the words students encounter as a tiered system. In their three-tiered system from Isabel Beck, Margaret McKeown, and Linda Kucan (2013), vocabulary words are classified into three tiers of organization.

- **Tier one** words are everyday, common words such as *more*, *less*, *high*, and *low*.
- **Tier two** words include vocabulary words that span across multiple academic settings such as *analyze*, *evaluate*, *explain*, and *justify*.
- **Tier three** vocabulary words are content-specific words like *quotient*, *quadratic*, *polynomial*, and *denominator*.

Literacy expert Kimberly Tyson (2013) suggests "understanding the three tiers can help separate the 'should-know words (tier three)' from the 'must-knows (tier two)' and the 'already-known words (tier one).'" Based on these tiers, the majority of your vocabulary instructional energy should focus on tier three with a check-in on tier two words, as needed.

Reflecting on Practice

In figure 3.1, Rheta N. Rubenstein and Denisse R. Thompson (2002) provide examples of "areas of challenge" that may make vocabulary development difficult for your mathematics students. After you review the chart and examples as a team, think about an upcoming unit of mathematics instruction. What are the challenges with the vocabulary your students will encounter throughout the unit? What words might students struggle with the most? (Hint: They might be words that explicitly appear in the standards and directly relate to the mathematics content you will teach [tier three vocabulary], but they might also be more global, cross-curricular words like *evaluate*, *analyze*, and *explain* [tier two vocabulary].)

Do you think you help your students reach clarity on vocabulary words? There are many informal words, symbols, and phrases used every day in mathematics classrooms, but in the end, that can be a sense-making detriment to students as mathematics content becomes

Area of Challenge	Possible Examples	Team Vocabulary Reflection for the Current Mathematics Unit
Mathematics and everyday English share some words, but they have different meanings in the two contexts or the mathematics meaning is more precise.	• *Right* angle versus *right* answer • *Foot* as twelve inches versus *foot* as body part • *Difference* as the answer to a subtraction problem versus *difference* as a general comparison	
Some mathematics words are found only in mathematical contexts.	• Quotient • Denominator • Integer • Isosceles • Histogram	
Some words have more than one mathematical meaning.	• *Square* as a shape versus *square* as a number times itself • *Round* as a shape versus *round* as a number operation	
Some mathematical words are related, but students may confuse their distinct meanings.	• *Hundreds* and *hundredths* • *Factor* and *multiple* • *At most* and *at least* • *Solve* and *simplify*	
English spelling and usage have many irregularities.	• *Four* has a u, but *forty* does not. • Fraction denominators, such as *sixth, fifth, fourth,* and *third,* are written like ordinal numbers, but rather than *second*, the next fraction is *half*.	
Some mathematics concepts are verbalized in more than one way.	• *Skip count* versus *find the multiples* • *One quarter* versus *one-fourth* • *Solutions*, *x-intercepts*, and *roots*	
Some mathematical words are homonyms with everyday English words.	• *Sum* versus *some* • *Arc* versus *ark* • *Pi* versus *pie* • *Graphed* versus *graft* • *Whole* versus *hole*	

Source: Adapted from Rubenstein & Thompson, 2002.

Figure 3.1: Team discussion tool—Categories of vocabulary challenges for students.

*Visit **go.SolutionTree.com/MathematicsatWork** for a free reproducible version of this figure.*

TEACHER *Reflection*

What words do you use with students on a daily basis for your most current unit of mathematics study? Which words or notations seem to consistently cause students problems during the unit?

How precise are you with your use of the academic language for each lesson, and how precise do you expect your students to be?

more complex. As you read the chart in figure 3.1, use the questions in the teacher reflection.

It's just as important for you, the teacher, to use correct mathematical terms and notation as it is for students to also use those terms correctly. For example, a high school teacher of mathematics who uses the word *corner* instead of *vertex* is not modeling the correct language that aligns to the standards. Neither is an elementary teacher who says the number is getting *bigger* instead of *increasing* in value. Your lack of precision can make it more difficult for students to mirror the same shifts in precision.

As students explain their reasoning and communicate with one another over more complex and conceptual mathematical tasks, their achievement does improve. "Knowledge of mathematics vocabulary affects achievement in mathematics, particularly in the area of problem solving," according to Robert Helwig and colleagues (Helwig, Rozek-Tedesco, Tindal, Heath, & Almond, 1999, p. 113).

John Hattie, Douglas Fisher, and Nancy Frey (2017) indicate "what we say to students, as well as how we say it, contributes to their identity and sense of agency, as well as to their success" (p. 203). Consider using the teacher reflection at right as a team activity as well, and

> **TEACHER** *Reflection*
>
> How do you currently establish academic vocabulary as part of the daily mathematics lesson every day?
>
> How do you encourage student communication that fosters accurate use of mathematics academic vocabulary and notation for all students, including your EL students?
>
> _____
> _____
> _____
> _____
> _____
> _____
> _____
> _____

discuss how you encourage classroom communication for vocabulary development.

If you do not address mathematics academic vocabulary as part of your daily lesson and overall unit design, students will lack agency—their voice—in learning. If

> *Personal Story* **JESSICA KANOLD-McINTYRE**
>
> In most schools I work with I hear a common theme when it comes to curriculum implementation. Teachers express concern that each teacher has his or her own language or words that he or she uses with students; teachers need help building a more consistent mathematics language in their classrooms. Every time I hear this concern, I am reminded that not only is it important for a classroom teacher to build vocabulary instruction into his or her lesson-design process, but it is equally important to discuss key vocabulary for the unit as a grade-level or course-based team so there are common expectations across classrooms in the school (K–5) or a mathematics department (middle school and high school). In order to ensure equity for all students in a grade level or a mathematics course, students need to experience accurate and common mathematics vocabulary, including notation.
>
> In my district's most recent curriculum writing process, we intentionally built in a place to record common academic vocabulary that we would all agree to teach for every mathematics unit. We created a systematic process in K–8 to ensure we were all in agreement on the common academic language vocabulary for each unit. This component of the unit-planning process also transferred to our middle school courses we had in common with the high school to ensure we had consistent expectations with the high school faculty as well.

you do not address academic vocabulary to equip all students with the confidence of using and understanding the language of mathematics as they discuss strategies to solve mathematics problems in class and embrace errors in their own thinking, then their learning will suffer during the lesson.

In part 2, chapter 9 of this book (p. 71), you and your colleagues will have a chance to explore how to implement academic language vocabulary strategies each day.

TEAM RECOMMENDATION

Include Academic Language Vocabulary as Part of Instruction

- In order for students to be able to justify, communicate, and reason mathematically, it is crucial they have a strong vocabulary of words, symbols, and notations.
- EL students need opportunities to learn mathematics vocabulary, engage in discourse, make conjectures, and explain their thinking (Civil & Turner, 2014).

Up to this point, the lesson-design elements have helped you to use the essential learning standards for the unit, identify the learning target for the lesson, determine *why* it is so important for students to learn, create a context for learning via the prior-knowledge warm-up task you choose, and finally consider the academic language vocabulary in the unit that can be a barrier to student learning. These lesson-design and team-planning actions help you to answer the first critical question of a PLC (DuFour et al., 2016), What do we want all students to know and be able to do? as well as, How can we help the students get ready to learn?

Armed with this wisdom, you and your colleagues now make the most important and powerful decision in your professional and daily work as a teacher of mathematics: you *choose* the mathematical tasks you will use to help your students demonstrate evidence of learning the new learning target before the lesson ends.

This is a lot of responsibility. Expert teachers ensure students will experience both higher-level-cognitive-demand tasks and lower-level-cognitive-demand tasks that align to the learning target for that day.

CHAPTER 4

Lower- and Higher-Level-Cognitive-Demand Mathematical Task Balance

> What should be the nature of mathematics that students learn—facts, skills and procedures or concepts and understanding? How should students learn mathematics—teacher directed with a focus on memorization, or student discovery through reasoning and discovery?
>
> —Philip S. Jones and Arthur F. Coxford

Although hard to believe, the preceding quote in the epigraph from Jones and Coxford was written in 1970. The debate between procedural knowledge and conceptual understanding in mathematics goes back a long time. Matthew R. Larson and Timothy D. Kanold (2016) establish that the debate surrounding the level of the cognitive demand of the mathematical tasks you choose to teach the lesson each day began in the 1820s and still exists today.

Furthermore, when it comes to mathematics task design, Larson and Kanold (2016) describe an equilibrium approach that balances the emphasis procedures and conceptual understanding as the best way forward for mathematics education.

As described so far in part 1, lesson design begins with the development of your team's understanding of the essential learning standards for the unit and the daily learning targets for each lesson in the unit. You also work to identify prior-knowledge standards and mathematical tasks with the academic vocabulary necessary to help your students persevere. These lesson-design and planning actions help you to answer the first critical question of a PLC for collaborative teams: What is it we want all students to know and be able to do?

And now, you choose the *mathematical tasks* you and your colleagues will use each day. There are few other decisions you make on a daily basis that have the same strength of impact on student learning as the choice of tasks you use for your lesson. The mathematical tasks and activities *you choose* for each lesson of the unit will help you to answer the second critical question of a PLC for collaborative teams: How will we know if they know it?

You choose the mathematical tasks for the lesson based on your judgment that those tasks will help your students to demonstrate an understanding of the learning target for the day. These task choices will also impact the rigor of the student learning experience. From an equity perspective, you want to select tasks characterized as "low threshold, high ceiling tasks" (McClure, Woodham, & Borthwick, 2011, p. 1) and that provide access and potential scaffolding entry points for all students. At the same time, the task you choose should have the potential to engage students in challenging mathematics and thinking at a much deeper reasoning level (Smith et al., 2017). Essentially a low-threshold, high-ceiling mathematics task is an activity where everyone in the student group can begin and work at his or her own level, yet the task also offers possibilities for learners or teams of student learners to do much more challenging mathematics using the task as well.

According to Melissa Boston and Margaret Smith (2009), a *mathematical task* is a single complex problem or a set of problems that focuses students' attention on a specific mathematical idea. Mathematical tasks include activities, examples, or problems that students complete as a whole class, in small groups, or individually. The tasks provide the rigor (levels of complex reasoning) you choose to determine the pathway of student learning, and to assess student success along that pathway during the lesson.

Additionally, a growing body of research links students' engagement in higher-level-cognitive-demand tasks to overall increases in mathematics learning, not just in the ability to solve problems (Hattie, 2012; Resnick, 2006).

Definition of Lower- and Higher-Level-Cognitive-Demand Tasks

There are several ways to label the cognitive demand or rigor of a mathematical task; however, for the purpose of this book and series, we classify tasks as either lower-level cognitive demand or higher-level cognitive demand as Margaret Smith and Mary Kay Stein (1998) define in their task-analysis guide that is printed in full as an appendix (page 117).

Lower-level-cognitive-demand tasks typically focus on memorization or rote procedures without attention to the properties that support those procedures (Smith & Stein, 2011). *Higher-level-cognitive-demand tasks* are tasks for which students do not have a set of predetermined procedures to follow to reach resolution, or, if the tasks involve procedures, they require that students justify why and how they perform the procedures.

Examples of lower-level- and higher-level-cognitive-demand tasks (either for use in class, for homework, or on a mathematics assessment) appear in figure 4.1 for various grade levels.

A key word regarding rigor is *balance*. Your choice of daily mathematical tasks for teaching the lesson should reveal a balance of procedural fluency and conceptual understanding proficiency throughout the lesson. Generally, the rigor-balance ratio is recommended to be about 50–50 (higher- to lower-level cognitive demand) during any unit, with the procedural knowledge and rote memorization student tasks *following* the development of conceptual understanding of the standards. This is a critical point: the procedural fluency type of mathematical tasks should be built on a foundation of conceptual understanding (NCTM, 2014).

Why Balancing Use of Lower- and Higher-Level-Cognitive-Demand Tasks Is Important

In order for students to develop true understanding of the essential learning standards, procedural fluency *and* conceptual understanding should be used in tandem (Kilpatrick, Swafford, & Findell, 2001):

> Procedural fluency and conceptual understanding are often seen as competing for attention in school mathematics. But pitting skill against understanding creates a false dichotomy. . . . Understanding makes learning skills easier, less susceptible to common errors, and less prone to forgetting. (p. 122)

Mathematics instruction should focus on developing skills and understanding for each student, as well as the ability for each student to reason with his or her understanding and skills to solve problems. Skills and understanding support one another, and are both necessary to be an effective problem solver (Larson & Kanold, 2016).

In the report *Education for Life and Work: Developing Transferable Knowledge and Skills in the 21st Century* (Pellegrino & Hilton, 2012), the expert panel cites research indicating deeper student learning of content ultimately produces retention and the ability to transfer knowledge learned at higher performance rates than does rote learning. "Rote learning of solutions to specific problems or problem-solving procedures" will not produce the competencies the 21st century requires (Pellegrino & Hilton, 2012, p. 9).

Directions: Choose the most appropriate grade level that follows, and, as a collaborative team, discuss why each of the questions meets the cognitive-demand levels to which it's assigned using the descriptions for lower- and higher-level-cognitive-demand tasks in the appendix (page 117).	
Grade 1: I can solve addition and subtraction word problems up to 20.	
Lower-level-cognitive-demand task: Add or subtract. 4 + 3 = ___ 6 − 2 = ___	**Higher-level-cognitive-demand task:** Isabel has 8 red flowers and 2 yellow flowers in a vase. How many flowers are in the vase altogether? Show how you know.
Grade 3: I can interpret whole-number quotients using objects.	
Lower-level-cognitive-demand task: Divide into groups of three. 	**Higher-level-cognitive-demand task:** Kyle sold some tubs of peanut butter cookie dough. Each tub costs $8. Kyle collected $32. How many tubs of peanut butter cookie dough did Kyle sell? Show your work and write an equation. Explain how you solved the problem.
Grade 6: I can write an inequality of the form $x < c$ or $x > c$.	
Lower-Level-Cognitive-Demand Task: Graph the inequality. $x \leq 14$ 	**Higher-Level-Cognitive-Demand Task:** Tony graphed $x < -4$ below. Is he right? Explain how you know. If he is wrong, correct his error.  −4
High school (algebra 2): I can simplify or solve rational expressions and equations.	
Lower-level-cognitive-demand task: Solve for x. Identify extraneous solutions. $\frac{2}{x} = \frac{x}{x^2 - 8}$	**Higher-level-cognitive-demand task:** If $\frac{2}{a-1} = \frac{4}{y}$ where $y \neq 0$ and $a \neq 1$, what is y in terms of a? For what values of a will y be negative? Show your work and justify your thinking.

Source: High school example adapted from the College Board, n.d.

Figure 4.1: Examples of lower- and higher-level-cognitive-demand mathematical tasks.

Personal Story SARAH SCHUHL

A teacher once told me the reason he loves teaching mathematics is because all he has to do is open the book each day to the lesson and show the examples from the text to students and assign homework. As I thought about his lesson-planning technique, I realized that he had a major misconception about teacher resources. How did he verify that the tasks in the publisher resources fully met the intent of the essential learning standard? Often the tasks in curriculum resources are in isolation from other lessons, focused only on the written learning target for that day's lesson, and they may or may not be a balance of lower- and higher-level tasks. In contrast, a teacher may need to combine content ideas, include specific process standards, or carefully choose which tasks to use because there may not be enough time to share and engage students in each task every day. In general, teams should ask, What is the purpose for using each mathematical task related to student learning during the lesson? What is the plan? Why should we use each task, and is there a balance in the cognitive-demand level of the tasks we choose?

In *Balancing the Equation*, authors Larson and Kanold (2016) state, "A modern definition of *mathematical literacy* includes student development of skills and procedures, conceptual understanding, problem solving, and a disposition to expend effort and persevere when learning mathematics and solving problems" (p. 62). This explanation establishes a working definition for the rigor you should seek in every mathematics lesson.

For example, it benefits students to use objects, number lines, skip counting, groups, and area models to build an understanding of multiplication before they can develop procedural fluency. Students need strategies to learn the multiplication facts beyond pure memorization with flash cards. An understanding of and development of flexible thinking with numbers is essential for elementary school students.

So, what are the sources you and your team can use to choose the mathematical tasks for your lessons each day? Start with publisher materials or previously used lessons to gather quality problems, questions, or mathematical tasks. From there, you may search websites for mathematical tasks and visit **go.SolutionTree.comMathematicsat Work** to access a set of free online resources, or look at tasks in other textbooks and resources as well.

Reflecting on Practice

Higher-level-cognitive-demand mathematical tasks are those that provide "opportunities for students to explain, describe, justify, compare, or assess; to make decisions and choices; to plan and formulate questions; to exhibit creativity; and to work with more than one representation in a meaningful way" (Silver, 2010, p. 2). In contrast, lessons or tasks with low cognitive demand are "characterized as opportunities for students to demonstrate routine applications of known procedures or to work with a complex assembly of routine subtasks or non-mathematical activities" (Silver, 2010, p. 2).

Thus, your teacher team should engage in meaningful dialogue and discussions about the choices of mathematical tasks members use to teach each lesson of the unit. You can use figure 4.2 to engage your team in these reflective conversations.

Once you include identifying higher-level-cognitive-demand mathematical tasks as part of your unit planning, you will begin to look at tasks and classify them as higher-level or lower-level cognitive demand and decide the best points of entry during the units' lessons.

Not all tasks need to be common across the team for each and every lesson within a unit; however, there needs to be consistency in the interpretation of the intended rigor of the essential learning standards and the expectations for proficiency. Even though the specific tasks you and your colleagues choose may not match exactly, the level of rigor students experience from one class to the next should be consistent. Also, the types of strategies and tools (manipulatives and others) students might utilize throughout the unit should be consistent. This ensures each and every student has access to rigorous learning opportunities.

As a team, identify and commit to the use of higher- and lower-level-cognitive-demand tasks on a consistent basis within your instruction. The balance of the types of tasks you choose supports student learning, encourages perseverance, and provides opportunities for you to truly assess students and respond to specific student needs.

How you select tasks and require students to engage in them matters. This selection establishes the expectation level for instruction and ensures that you engage each student in higher-level-cognitive-demand tasks, providing the opportunity for students to go deep with mathematics, one of the equity-based mathematics teaching practices (Aguirre et al., 2013).

TEAM RECOMMENDATION

Examine Your Cognitive-Demand Task Balance

- Mathematical tasks can be a single problem or a set of problems and activities you use to develop student understanding for each essential learning standard of the unit.

- It's important to use a balance of lower- and higher-level-cognitive-demand tasks within the daily lessons of the unit in order to support a balance of student development in procedural fluency and conceptual understanding.

- Higher-level-cognitive-demand tasks support increased student exploration, communication, and reasoning and allow for multiple solution pathways or representations.

Directions: As a team, use the following questions to discuss how you currently select and use higher- and lower-level-cognitive-demand tasks within your lesson-design process.

1. Describe some of your favorite mathematics problems to use during this unit and how you use them to teach the corresponding essential learning standard.

2. How do you define and differentiate between higher-level-cognitive-demand and lower-level-cognitive-demand tasks for each essential learning standard of the unit?

3. What percentage of your current mathematics tasks you use during instruction fall into the lower-level-cognitive-demand category, and what percentage fall into the higher-level-cognitive-demand category? (Provide an average.)

4. How do you work as a team to select specific common higher-level-cognitive-demand and lower-level-cognitive-demand mathematics tasks that all students of the grade level or course will experience for each essential standard of the unit?

5. Does your team have a proper balance of mathematics tasks you present to students throughout the unit of instruction in terms of the complexity of student reasoning the tasks require? Please explain.

6. How might what you learn about your students' understanding of the essential learning standard differ depending on the cognitive demand of the mathematical tasks you use during instruction?

7. How do you use higher-level tasks to provide feedback to individual students and groups of students during the lesson?

Figure 4.2: Team discussion tool—Choosing mathematical tasks for lesson design during the unit.

Visit go.SolutionTree.com/MathematicsatWork for a free reproducible version of this figure.

So far in part 1, you have examined lesson-design elements to help you identify daily learning targets for each essential standard for the lesson and determine why those standards are important, created a context for learning via the prior-knowledge warm-up task you choose, considered how the academic language vocabulary in the unit can be a barrier to student understanding and learning, and discovered the expectation to use a balance of higher- and lower-level-cognitive-demand mathematical tasks throughout the unit. These lesson-design and planning actions serve the first two critical questions of a PLC (DuFour et al., 2016), What do we want all students to know and be able to do? and, How will we know if they learn it?

In order to implement your chosen mathematical tasks for each lesson well, your lesson planning and design need to next consider carefully the type of discourse students will use to engage in each mathematical task. In the next chapter, you explore whole-group versus small-group discourse activities and how to balance between the two types of student communication during the lesson.

CHAPTER 5

Whole-Group and Small-Group Discourse Balance

Tell me and I forget, teach me and I may remember, involve me and I learn.

—*Benjamin Franklin*

The nature of the classroom discourse during the lesson is the next logical choice to consider as you design a mathematics lesson.

A primary purpose of the lesson is to *actively engage all students* in demonstrating evidence of learning the essential learning standard and more specifically the learning target for that day. In order to honor this purpose, you will need to carefully consider the nature of the student discourse during the lesson.

TEACHER *Reflection*

Where and when are student peer-to-peer conversations currently taking place as part of your mathematics lessons?

Who is doing most of the talking throughout a lesson—you to your students, or your students with one another?

_____
_____
_____
_____
_____
_____
_____

Facilitating Discourse in the Classroom

There are generally two primary types of discourse as you plan the strategies and activities or actively engage your students in conversations with each task of the lesson.

1. **Whole-group discourse:** The teacher leads the discussions and models how to do a problem. He or she calls on students one at a time to respond to questions as the task solution unfolds and reflects student responses back to other members of the class.

2. **Small-group discourse:** Students work together with a time limit and specific directions or prompts provided for expected discussions and sharing. Students collaborate with other peers and process through strategies together for an assigned mathematics task or discussion prompt.

The manner in which you facilitate student discourse is essential in creating and supporting a classroom-learning environment that values reasoning and sense making from the *student's point of view*. How you ask questions and facilitate discourse in your classroom has important implications for whether or not your classroom instruction promotes an equity culture. The positioning of your students influences how students see themselves as members of the community (Smith et al., 2017).

Margaret Smith, Michael Steele, and Mary Lynn Raith (2017) offer the following additional lesson discourse reflection questions:

- Are all students' ideas and questions heard, valued, and pursued in the mathematics classroom?
- What mathematical ideas does the class examine and discuss?
- Whose thinking does the teacher select for further inquiry, and whose thinking does the teacher disregard during small-group and whole-class discussion?
- Who in the classroom is positioned as competent?
- Whose ideas are featured and privileged? (p. 95)

NCTM (1991, 2014) highlights the importance of classroom discourse for almost three decades: "Teachers, through the ways in which they orchestrate discourse, convey messages about whose knowledge and ways of thinking and knowing are valued, who is considered able to contribute, and who has status in the group" (p. 20).

Simply put, the discourse interactions in the classroom have profound implications for how students see themselves. Your interactions:

> Must help students see themselves as people who can know, do, and make sense of mathematics, challenging aspects of marginality that lie within students' own identities. The language choices that teachers and students make, the norms that are set in the classroom, and the types of mathematical interactions that are or are not encouraged can all support or inhibit the development of productive attitudes and practices toward mathematics. (Smith et al., 2017, p. 140)

Through the use of either whole-group or small-group discourse, you decide what thinking to share and whose voices to hear. This has a profound impact on how knowledge is shared and created in your classroom, how deeply the task engages students, how they will receive feedback during the task, and who plays a primary role in that knowledge sharing and creation. In short, your discourse affects student agency toward learning the mathematics. Will the students persevere during the lesson and the tasks you have prepared? What if they get stuck while working on the tasks you have chosen? Then what?

Thus, part of your lesson-design plan is to determine *how* your students will experience the process of learning the mathematical tasks you chose.

- Will students watch you model the task, as you ask individual members of the class guiding questions from the front of the room? (Whole-group discourse)
- Will your students discuss a strategy for working on the task with partners or in small-group discussions as you circulate the room and provide differentiated feedback and prompts for perseverance? (Small-group discourse)
- Or, will students use some combination of both lesson communication styles?

As a teacher, you make these decisions on a daily basis in your lesson-planning process. It is important to note that there are severe limitations to using whole-group discourse *only*.

As they relate to student perseverance in class, mind wandering and *cascading inattention*—overload in the mind—are due to the complexity of information you deliver to the students during the whole-group phase of a lesson. As John Hattie and Gregory Yates (2014) indicate, "Student attention deteriorates over the course of a lesson" during whole-group discourse (p. 39).

Thus, too much time spent using whole-group discourse instruction can be very damaging to student learning. The National Board for Professional Teaching Standards (2010), in its mathematics report, states it more succinctly: "In an environment of trust, students feel safe to communicate different points of view, to conduct open-ended explorations, to make mistakes, and to admit confusion or uncertainty in order to learn" (p. 36).

Hattie and Yates (2014) further indicate the effect size on student learning is significant (0.82 standard deviations above the norm) when students in your class see each other as reliable and valuable resources for each other, and when they express their ideas, questions, insights, and difficulties *student to student*.

Small-group discourse is all about what you *see* and *hear* the students thinking and reasoning about with one another. Part of your professional lesson-design responsibility is to ensure a daily balance of whole-group and small-group classroom discourse or discussions.

Remember, the intent of your lesson design is to be able to answer PLC critical questions 1 and 2 (DuFour et al., 2016) for your students during the lesson.

1. What do we want all students to know and be able to do?
2. How will we know if they learn it?

To a great extent, if most of your lesson design is to use whole-group discourse from the front of the classroom, you will not be able to determine an answer to PLC critical question 2. You will only know the answer for the few students you call on.

On the other hand, student small-group discourse opens a window into student reasoning about mathematics. You *hear and see* the connections students make, the questions they ask, as well as the obstacles or misconceptions that can hinder their conceptual understanding.

It is mostly during the small-group discourse moments of the lesson that you can get a more accurate assessment concerning the essence of PLC critical question 2, How will we know if they are learning the standard? By listening carefully to students you are in a better position to make decisions and implement strategies to support learning and push the level of reasoning and problem solving you expect to a higher level.

The subsequent small-group student dialogue creates a community of learners who collectively build mathematical knowledge and proficiencies and *persevere* together. More important, as you will also discover in part 2, chapter 10 of this book (page 77), you will improve your ability to effectively answer PLC critical question 3 (DuFour et al., 2016), How will you respond when some students do not learn (during small-group discourse activities)? As your students make initial errors in their reasoning and thinking on the tasks you present to them, you are able to provide more accurate and meaningful student feedback to the lessons' mathematical tasks.

Thus, you deliberately structure opportunities for students to use and develop appropriate mathematical discourse as they reason and solve problems with one another. This means you give students opportunities to talk with one another, work together in solving problems, and discuss their mathematical thinking and understanding with their peers.

Using Whole-Group Discourse

Does this mean that whole-group discourse with teacher lecture or modeling from the front of the room is a bad practice? No. But it is an overused teacher practice currently that severely limits student learning if you were to rely on it too often and too narrowly. As the title of this chapter of part 1 suggests, your goal in planning is to seek *balanced student communication and discourse* around the tasks you choose for the lesson.

There are different situations where whole-group activities make sense. Three are as follows.

1. When you are introducing a new standard or modeling a standard with deeper connections and explanations, you most likely need to allow students to observe first, and then comment together and reflect upon your modeling and thought process.

2. When you are leading the class through a mathematics task, you can occasionally blend some student partner talk around a question or prompt arising from the nuances of the mathematical task you are asking students to think through. You can, as Mike Schmoker (2011) recommends, transform a traditional teacher-led lecture into an interactive lecture that raises student engagement. You do not really yield the whole-group discourse completely, as you give your students brief moments of discussion time. You can keep the attention to the overall mathematics task flowing through your advice and modeling. This teacher action engages the students on a much deeper level than just calling on one or two students at a time during your whole-group moments of the lesson.

3. Sometimes, once you complete a small-group activity, it may help if you provide a summary of an important reveal from the small-group discourse learning experience, make mathematical connections students may have missed, or draw out the mathematical structure (Smith & Stein, 2011).

One caution, though, is to rarely "go over" with the whole group a problem the students just worked on during small-group discourse. If you do, students will learn quickly to not persevere on the task you are asking them to discuss with their peers, as they can just wait it out knowing you will show them how to do the mathematics task in just a few minutes.

Using Small-Group Discourse

In order to balance whole-group with small-group activities at different times within a lesson, your students should be sitting in groups or teams—preferably teams of four.

By setting up a culture of collaboration with your students, the transition between whole-group activities and small-group activities during the lesson becomes seamless. In order to facilitate both types of discourse well, it is necessary to use clear small-group and whole-group structures and expectations. You will explore *how* to use these structures for a more formative feedback purpose in part 2, chapter 10 of this book (page 77).

Small-group discourse for teaching mathematics is a powerful lesson tool, and you will need to plan for it accordingly. When author Jessica Kanold-McIntyre observes colleagues during a mathematics lesson, she notes that the most successful colleagues are able to *establish clear norms and expectations* for how students are to participate together during the small-group discourse and within a clearly allotted amount of time.

Remember that either extreme for the planning of the lesson discourse is not good. Too much time in small-group discourse can be just as damaging to student learning as too much time lecturing from the front of the room. The goal is to seek balance and to embrace the value of whole-group discourse as you need it.

Personal Story TIMOTHY KANOLD

By 1994, after eight years of working on our vision for mathematics instruction, our teachers at Adlai E. Stevenson were frustrated by the lack of authentic student engagement in our lessons. Our teachers felt that one of the primary barriers to student learning was our traditional acceptance of students sitting in rows. We had done an exhaustive review of the research and could not find any research that students sitting in rows would improve student achievement in any subject area—much less mathematics. One of our younger math teachers, Karen O'Staffe, asked, "Why do the students have to sit in rows?", and our small team of teachers investigating this issue agreed: "They don't." For our faculty, the mathematics instructional revolution finally began, and the results of our student mathematics performance began to soar.

We knew our expected instructional model for mathematics could no longer accommodate students sitting in rows. We had this deep-seated awareness that our students were just not engaged in the lesson due to our constant reliance on lecture or whole-group discourse.

As we were shaping our early work as part of the PLC process, we decided to pilot a different classroom desk arrangement structure using heterogeneously mixed groups of students who would be taught how to do work together in meaningful discussions about potential solution pathways to the work we were presenting. We called it "teams of four."

The problem, of course, was that we had no idea how to manage the student teams and how to help them effectively engage. So we met together every Wednesday afternoon to share what was working, what was not working, and how to improve our expectations for the students to become a community of learners. Over time, our teachers—and our students—began to embrace a new culture for student learning.

Personal Story JESSICA KANOLD-McINTYRE

In my early years of teaching middle school mathematics and now with my experience as a principal observing mathematics instruction, small-group activities are one of the most challenging aspects of lesson design. There were three types of barriers I faced during my early years of teaching that I also noticed in the mathematics instruction throughout our school.

First, small-group discourse with a mathematics task at times can cause feelings of losing control. To be successful at using small-group discourse, I became purposeful and intentional at planning to set up my expectations for how I expected my middle school students to work and share. These directed prompts involved both time and specificity in terms of my expectations for student communication during the mathematics task.

Second, the most accomplished mathematics teachers I have observed are able to use small-group discourse while preventing student conversations from veering off topic. These teachers have assessing and advancing prompts (see part 2, page 77 for more information) in place for orchestrating the small-group discourse as he or she monitors the room and provides engaging feedback prompts to the students.

Third, it is helpful to use clear transitions during the lesson between whole-group and small-group discourse activities. Various strategies, prompts, and structures help students persevere during small-group discourse and for the transitions back to whole-group discourse. These activities will be discussed in more detail in part 2 (page 77).

From the front of the room, your opportunity to monitor and provide meaningful formative feedback to each of your students decreases. You are only one teacher for many students—sometimes thirty or more. However, if you utilize small-group discourse interwoven into your lessons, you can use eight student groups of four in a class of thirty-two for feedback and action on the mathematical tasks of the lesson. It is much easier to circle around and give meaningful feedback to eight teams of students as they reason through a mathematical task, rather than more than thirty students individually.

Reflecting on Practice

As a team, use figure 5.1 (page 44) to respond to questions about your current small-group or whole-group student practice and discourse.

Complete the teacher reflection (page 45) once you respond to the questions from figure 5.1.

To this point in part 1, your daily mathematics lesson design includes identifying the essential learning standard for the unit and the related daily learning target for the lesson, identifying the prior knowledge necessary for student success, and determining the academic language and vocabulary that need to be included as part of your instruction within the lesson. You've also examined the importance of carefully choosing the mathematics tasks that you believe will support student learning of the standard through balancing the cognitive-demand level of those tasks.

Finally, this chapter has shown that as you choose the various tasks for your lesson, you then must choose the nature of the student communication and discourse for each mathematics task, carefully balancing whole-group and small-group discourse.

So now that the lesson is about to end, what do you do next? Surprisingly, in many K–12 mathematics lessons, the answer is nothing. You are simply out of time. Yet just as you should warm students up at the start of the lesson with a prior-knowledge task or prompt, you should also help them cool down. Lesson closure is the focus of the next chapter, and the final lesson-design element.

Directions: Think about the nature of student discourse, generally, during your current mathematics lessons.

1. Consider the mathematics tasks you currently use to teach an essential learning standard. What percent of the lesson time do you spend using whole-group discourse versus small-group discourse as students work through and learn from the solutions? Is the type of discourse your students experience balanced?

2. What type of structures do you like best for small-group student discussions? Teams of four? Partners? Threes? Explain why.

3. What is your biggest obstacle to including more opportunities for peer-to-peer small-group discourse time into your lessons? What can you do to remove that obstacle?

4. Do you utilize small-group discourse during your review activities such as warm-ups or preparation for an upcoming assessment? If so, share what works best for you.

5. What is your best advice to ensure students persevere and stay engaged during whole-group discourse activities or direct instruction?

6. How do you ensure each student answers questions during a lesson and engages in learning through the tasks for both whole-group and small-group portions of the lesson?

Figure 5.1: Team discussion tool—Student discourse during the lesson.

*Visit **go.SolutionTree.com/MathematicsatWork** for a free reproducible version of this figure.*

TEACHER *Reflection*

As you reflect on your answers to the questions in figure 5.1, list two or three lesson-design teaching practices you are now thinking about differently.

What actions in regard to student discourse, whole group or small group, might you be able to implement in your classroom tomorrow?

_____

_____

_____

_____

_____

_____

TEAM RECOMMENDATION

Balance Whole-Group and Small-Group Discourse Activities

- In order to balance small-group and whole-group discourse, it's beneficial for students to be sitting in groups of four or some other team arrangement.

- Your best opportunity to see and hear how students are thinking, monitor that student thinking and learning, and provide formative feedback occurs when students are working in small groups on mathematical tasks with their peers as you monitor the student teams to see and hear what they are doing and thinking.

CHAPTER 6

Lesson Closure for Evidence of Learning

> We need to shift towards classrooms as mathematical communities . . . away from classrooms as simply a collection of individuals.
>
> —National Council of Teachers of Mathematics

How do you know if your lesson achieves its purpose? How do you know if your students can demonstrate evidence of learning the standard for the lesson? How do students know if they meet the learning target for the day? A lesson-closure activity allows students to summarize their learning from the lesson and for you to determine the class's progress toward the essential learning standard.

Hattie (2009) references closure as follows:

> [As a way] to cue students to the fact that they have arrived at an important point in the lesson or the end of a lesson, to help organize student learning, to help form a coherent picture, to consolidate, eliminate confusion and frustration, and so on, and to reinforce the major points to be learned. Thus closure involves reviewing and clarifying the key points of a lesson, tying them together into a coherent whole, and ensuring they will be applied by the student by ensuring they have become part of the student's conceptual network. (p. 33)

Student-led summaries of learning are the best way for you to obtain information on student progress.

Student-Led Summaries

A *student-led summary* is the closure activity you choose to engage your students with in a "cool down" at the end of the lesson. The benefit and purpose of a student-led summary are that it allows students to take ownership of their own learning while also providing you with feedback about what your students actually learned. This type of closure activity is a review prompt of the content from the lesson or learning target for the day.

Student-led closure is different than a final check for understanding, such as an exit slip or solving a couple of mathematics problems on a sheet of paper before wrapping up the lesson. A student-led summary can occur before or after the exit slip, but it needs to reveal how the student *understands* and has processed the lesson activities.

The end of a lesson provides an opportunity to ask questions and plan for activities that tie together the content of the lesson. The focus of this time should be to help your students make connections and help them to address misconceptions you have observed during the lesson. It can also reveal how they feel about the lesson, and their level of confidence for understanding the standard.

As a team, share different ways that each of you plan for and enact lesson closure. This is a great time to learn from each other and share ideas. Use the following team discussion tool (figure 6.1, page 48) to support these professional conversations.

As you choose the way you will close the lesson, be sure to connect the discussion prompts you provide back to the learning target or targets for the day, and the essential learning standard for the unit. The closing question, prompt, or activity you choose should also support an appropriate way for students to have time to reflect on the overall lesson experiences so they can accurately summarize and connect their learning.

Directions: As a team, answer each of the following questions.

1. What are some ways you currently embed student-led closure activities? If you don't have many student-led closure activities, how could you adjust current closing activities to make them student led?

2. How do you use the information from the student-led closure as part of a formative process with feedback on student responses? What evidence are you collecting during closure, and how do you respond?

Figure 6.1: Team discussion tool—Lesson-closure reflection.

*Visit **go.SolutionTree.com/MathematicsatWork** for a free reproducible version of this figure.*

Personal Story 66 TIMOTHY KANOLD

At Stevenson, we had a mathematics teacher, Mary Layco, who would eventually become recognized as an outstanding teacher in Illinois. The first time I watched her teach a math lesson, she passed out paper plates to every student and had them stand in circles with five students per circle. This was with about seven minutes left to go in the lesson. She asked the students to write down in the middle of the plate their answer to this prompt: "What is one thing you would like to know more about from today's math lesson?"

She then asked them to rotate the plates one person to the left and provide an answer to the question on the plate. After five rotations (about one minute each), each plate was back to the original owner with five answers to the question! I remember thinking, "Wow! What a great idea!" and immediately used the strategy with my students. She did not know it at the time, but Mary was creating intentional student-led summaries of her lesson.

Sample Closing Prompts

Detailed strategies for closing prompts appear in part 2 of this book (chapter 11, page 99); however, following are some sample closure prompts or questions.

- "How did mathematics tasks _____ help you understand the learning target _____ today?"
- "What is one thing you understand about [*learning target*] that you did not know this morning?"
- "How would you explain [*learning target*] to your friend if he or she were absent today?"
- "What is one thing you would like to know more about based on today's lesson?"
- "On a piece of paper or in your journal, summarize the lesson in three sentences."
- "List in your notebook the one thing that made the most sense to you today."
- "When someone asks you what you learned in math today, what would you tell them? Start with 'I can . . .'"
- "What part of the math lesson was most confusing for you today? State it in one sentence."
- "Based on this unit so far, name the math standard you would like to practice more before the unit ends."
- "Go to your online Google form and work in pairs to describe what you learned today and your confidence in doing the mathematics."

Remember, closure to the mathematics lesson is a quick review to connect your students to what it was they have learned (the daily learning target) and allows you to gather information on their understanding and progress as you plan the next lesson forward.

Furthermore, closure promotes clarity to your mathematics lesson, as you bookend your opening (Why are we learning to do these math problems today?) to your closing (Does the student have clarity on what he or she actually learned?). This, in turn, promotes the idea of student agency as part of the daily owning of their learning, and their understanding of what they were to have learned.

Reflecting on Practice

Reflect for a few moments on these closure prompts you use as part of your current lessons in the teacher reflection.

TEACHER *Reflection*

How do you currently plan for student-led lesson-closure activities as part of your lesson each day?

What is your favorite closure activity? List different strategies and ideas.

_____
_____
_____
_____
_____
_____
_____
_____
_____

The academic work of analyzing and summarizing what has been learned as part of the lesson each day should be done by your students—not by you. Closure allows your students to summarize main ideas, potentially answer questions posed at the beginning of the lesson, and help your students demonstrate their own meaning-making. How did the lesson turn out from *their* point of view—not your point of view?

TEAM RECOMMENDATION

Use Student-Led Closure

- Student-led summaries of learning are the best way for you to gain information on student progress and confidence at the end of a lesson.
- Student-led summaries should be connected to the daily learning targets and essential learning standard for the unit.
- Be sure to allow an appropriate way for students to reflect regarding their overall lesson experience as they make connections to their learning experience for that day.

This design element is the final step in your design process for high-quality lessons. Part 1 has included identifying the essential learning standard for the unit and the related daily learning target for the lesson, identifying the prior knowledge necessary for student success, and determining the academic language and vocabulary that you need to include as part of your instruction within the lesson. It also focused on the importance of carefully choosing the mathematics tasks that you believe will support student learning of the standard and that these tasks are balanced cognitively. Then, you learned about the need to choose the various tasks for your lesson and about choosing the nature of the communication and discourse for each task, carefully balancing whole-group and small-group discourse.

In the next chapter, you will study and review a sample lesson-planning template for mathematics that incorporates all six of the lesson-design elements described in part 1.

CHAPTER 7

Mathematics in a PLC at Work Lesson-Design Tool

Recall there are four critical questions (DuFour et al., 2016) every collaborative team in a PLC asks and answers on a unit-by-unit, ongoing basis.

1. What do we want all students to know and be able to do? (The learning targets)
2. How will we know if they learn it? (The assessment instruments and tasks)
3. How will we respond when some students do not learn? (Using a formative assessment process)
4. How will we extend the learning for students who are already proficient? (Using a formative assessment process)

Figure 7.1 (pages 52–53) presents the Mathematics in a PLC at Work lesson-design tool that helps ensure your teacher team reaches mathematics lesson clarity on all four of these PLC critical questions.

Planning With Your Team

Effective mathematics instruction rests in part on your careful planning (Morris, Hiebert, & Spitzer, 2009). Your collaborative team is uniquely structured to provide the time and support it needs to interpret the standards, embed student-engaged discourse practices into daily lessons, and reflect together on the effectiveness of your implementation, sharing evidence of student learning.

You can use the design tool during your planning for the instruction of the mathematics lesson as a way to support the focus and design of the mathematics tasks you choose to teach each learning standard.

Your collaborative team is most likely using some type of lesson-planning format or tool currently. What separates the mathematics lesson-design tool in this chapter from most other lesson-planning models is the focus on directly planning tasks from the student perspective in order to support students in reaching proficiency on the mathematics content standards of the lesson.

As you review this lesson-planning template, you will observe the six lesson-design elements part 1 explores.

1. Essential learning standards: the *why* of the lesson
2. Prior-knowledge warm-up activities
3. Academic language vocabulary as part of instruction
4. Lower- and higher-level-cognitive-demand task balance
5. Whole-group and small-group discourse balance
6. Lesson closure for evidence of learning

There are also a few additional planning components that will contribute positively to your planning process.

First, under the essential learning standard, there is a place to list the *content* essential learning standard. However, there is also a place to list the *process* learning standard. That references how you hope your students process learning the content standard during the lesson. Reference the NCTM process standards at www.nctm.org/Standards-and-Positions/Principles-and-Standards/Process, your specific state or province's mathematical practice standards, or the mathematical proficiencies in the National Research Council's book *Adding It Up* (Kilpatrick et al., 2001).

Second, the Mathematics in a PLC at Work lesson-design tool also specifically asks you and your team to consider both student and teacher actions during lesson planning. It is designed to help you to "see" learning through the eyes of students and help them become owners of their learning.

Preparing for the Lesson
Unit: Fill in the title of the unit. **Date:** Fill in the date of the lesson. **Lesson:** Provide a short descriptor about the nature of this lesson.
Essential learning standard: State the essential content and process standard *for the unit* you address during *this* lesson. • **Content**—Write as an *I can* statement. • **Process**—Write as an *I can* statement.
Learning target: State the specific learning outcome(s) for this lesson. *Use, "Students will be able to . . ."*
Academic language vocabulary: State the academic vocabulary expectations for the lesson. Describe how you will explicitly address any new vocabulary.

Beginning-of-Class Routines
Prior knowledge: Describe the warm-up activity you will use. How does the warm-up activity connect to students' prior knowledge, connect to an analysis of homework progress, or connect to future learning?

During-Class Routines
Task 1: Cognitive Demand (Circle one) *High* or *Low* What are the learning activities to engage students in learning the target? Be sure to list materials as necessary.

What will you be doing?	What will the students be doing?
• How will you present and then monitor student response to the task? • How will you expect students to demonstrate proficiency of the learning target during in-class checks for understanding? • How will you scaffold instruction for students who are stuck during the lesson or the lesson tasks (assessing questions)? • How will you further learning for students who are ready to advance beyond the standard during class (advancing questions)?	• How will you actively engage students in each part of the lesson? • What type of student discourse does this task require—whole group or small group? • What mathematical thinking (reasoning, problem solving, or justification) are students developing during this task?

Task 2: Cognitive Demand (Circle one): *High* or *Low*
What are the learning activities to engage students in learning the target? Be sure to list materials as necessary.

What will the teacher be doing?	**What will the students be doing?**
• How will you present and then monitor student response to the task? • How will you expect students to demonstrate proficiency of the learning target during in-class checks for understanding? • How will you scaffold instruction for students who are stuck during the lesson or the lesson tasks (assessing questions)? • How will you further learning for students who are ready to advance beyond the standard during class (advancing questions)?	• How will you actively engage students in each part of the lesson? • What type of student discourse does this task require—whole group or small group? • What mathematical thinking (reasoning, problem solving, or justification) are students developing during this task?

Task 3: Cognitive Demand (Circle one): *High* or *Low*
What are the learning activities to engage students in learning the target? Be sure to list materials as necessary.

What will the teacher be doing?	**What will the students be doing?**
• How will you present and then monitor student response to the task? • How will you expect students to demonstrate proficiency of the learning target during in-class checks for understanding? • How will you scaffold instruction for students who are stuck during the lesson or the lesson tasks (assessing questions)? • How will you further learning for students who are ready to advance beyond the standard during class (advancing questions)?	• How will students be actively engaged in each part of the lesson? • What type of student discourse does this task require—whole group or small group? • What mathematical thinking (reasoning, problem solving, or justification) are students developing during this task?

End-of-Class Routines

Common homework: Describe the independent practice teachers will assign when the lesson is complete.

Lesson closure for evidence of learning: How will lesson closure include a student-led summary? By the end of the lesson, how will you measure student proficiency and that students develop a deepened (and conceptual) understanding of the learning target or targets for the lesson?

Teacher end-of-lesson reflection: (To be completed by the teacher after the lesson is over)
Which aspects of the lesson (tasks or teacher or student actions) led to student understanding of the learning target? What were common misconceptions or challenges with understanding, if any? How should you address these in the next lessons?

Figure 7.1: Mathematics in a PLC at Work lesson-design tool.

*Visit **go.SolutionTree.com/MathematicsatWork** for a free reproducible version of this figure.*

Third, as we describe in more detail in part 2 of this book, the Mathematics in a PLC at Work lesson-design tool will help you plan for formative assessment feedback questions and student evidence of learning during the mathematics lesson. These communication choices include:

- How do you and your colleagues expect students to express their ideas, questions, insights, and difficulties?

- Where, when, and between whom should the most significant conversations be taking place (student to teacher, student to student, or teacher to student)?

- How do you use questions to facilitate small-group peer-to-peer discourse and differentiate learning around higher-level-cognitive-demand tasks during the lesson?

- How approachable and encouraging should you be as students explore? Do students see each other as reliable and valuable learning resources, as well as you?

Author Matt Larson reveals in his story ways in which you can make your learning more visible to each other as you help one another observe for evidence of the Mathematics in a PLC at Work lesson-design tool criteria in your own lesson-planning and implementation efforts.

Personal Story MATT LARSON

A principal asked me to observe and provide feedback concerning the classroom environment of one of her mathematics teachers. The principal was concerned because each semester a number of students asked to transfer out of the teacher's classes. All outward signs were the teacher was highly effective. There were no discipline issues, and his colleagues valued his participation on their collaborative team where he made significant contributions to task selection and assessment design.

I spent a day observing the teacher, and it was quickly clear what he could address to improve his classroom environment: his questioning pattern. The teacher's questioning pattern was to pose a question, leave little wait time, call on students, and then evaluate their response. His questioning pattern communicated to students that all that mattered were quick and correct answers. In addition, if students answered incorrectly, they felt demeaned by his response to them. Students simply disengaged.

He agreed to videotape his classroom, and we analyzed the tape together. We focused on the impact of his questioning pattern on how students saw themselves as learners. We worked together on his questioning approach. Over the course of the semester, he began to ask questions and listen to student responses to determine students' level of understanding, have other students respond to student responses, show respect for students' responses, and ask more open-ended questions that provided space for students' ideas; in short, he began to position students as capable doers of mathematics and, in the process, he completely transformed his classroom environment.

Thus, learning becomes more *visible* to you, your colleagues, and your students. One particularly powerful collaborative tool is lesson study. Research shows that lesson study is very effective as a collaborative protocol with a high impact on teacher professional learning (Gersten, Taylor, Keys, Rolfhus, & Newman-Gonchar, 2014; Stigler & Hiebert, 1999). Your team can also use the tool as you participate in collective team inquiry during the unit. *Mathematics Coaching and Collaboration in a PLC at Work* addresses these mathematics lesson-study protocols and methodology in great detail.

Reflecting on Practice

As a team, use figure 7.2 to reflect on the elements of the Mathematics in a PLC at Work lesson-design tool as you bring your own personal closure to part 1 of this book.

TEAM RECOMMENDATION

Use and Reflect on the Mathematics in a PLC at Work Lesson-Design Tool

- Collaborate as a team to collectively write common essential learning standards for the unit, daily learning targets for each lesson, and vocabulary for every lesson as well.
- Collaborate and share ideas about prior-knowledge warm-up activities, mathematics task selection, whole-group and small-group discourse activities, and student-led closure.

Directions: Closely examine the components in the Mathematics in a PLC at Work lesson-design tool (figure 7.1, pages 52–53). With your colleagues, discuss how you could use parts or all of the tool to help your current efforts to plan mathematics lessons for your grade level or course.

Lesson-Design Element	Strengths Identify what you currently do that is a strength for each component.	Challenges Identify any components that you currently do not address or do not address well in your lesson-planning process, and list how you might improve these components.
1. Essential learning standards: the *why* of the lesson		
2. Prior-knowledge warm-up activities		
3. Academic language vocabulary as part of instruction		
4. Lower- and higher-level-cognitive-demand mathematical task balance		
5. Whole-group and small-group discourse balance		
6. Lesson closure for evidence of learning		

Figure 7.2: Team discussion tool—Protocol for team analysis of the Mathematics in a PLC at Work lesson-design tool.

*Visit **go.SolutionTree.com/MathematicsatWork** for a free reproducible version of this figure.*

PART 1 SUMMARY

In part 1, you and your team examined the *what* of the six essential lesson-design elements for student success.

1. Essential learning standards: the *why* of the lesson
2. Prior-knowledge warm-up activities
3. Academic language vocabulary as part of instruction
4. Lower- and higher-level-cognitive-demand mathematical task balance
5. Whole-group and small-group discourse balance
6. Lesson closure for evidence of learning

Effectively implementing the lesson-design elements supports equitable teaching practices and outcomes. In fact, the concept of team action, the foundation of this book and the other books in the *Every Student Can Learn Mathematics* series, makes equity in learning possible by focusing the work of your collaborative team on the four critical questions of a PLC (DuFour et al., 2016). Thus, teams establish goals to focus learning and ensure that all students have access to the learning—regardless of the teacher students receive.

The goals you and your team select for your students have a profound impact on their development of a productive mathematics disposition (Smith et al., 2017). "The teacher's goals for a lesson frame the tasks that teachers choose and therefore students' opportunities to learn" (Smith et al., 2017, p. 26).

As you and your team plan lessons, you may find that you have different ideas or strategies for how to plan or teach a specific standard. This is okay. When it comes to lesson planning and the lesson-design process, it is important to ensure everyone is clear and consistent on the use of the essential learning standards and the academic language vocabulary as well as the level of rigor each standard requires in your selection of lower-level and higher-level tasks. However, there may be variation in the prior-knowledge, discourse, and closure activities you choose due to the formative nature of these actions; you may need to choose different activities based on evidence of learning (or lack of evidence) your students demonstrate throughout instruction.

To ensure consistency in the interpretation of the essential learning standards in the unit, you and your team can use the tool in figure P1.3 (pages 58–59) to start a conversation at the beginning of the unit about each of the specific lesson-design elements you consider as you each write your lessons.

As a team, use figure P1.3 at the beginning of a unit to help organize your team planning. List all of the essential learning standards for the unit and common academic language vocabulary, and, as a team, discuss ideas you may have for prior-knowledge warm-up tasks, different tasks you plan to use, and ideas you could use to implement lesson closure for evidence of student learning for each of the essential learning standards. As you fill out each component be sure to discuss how you plan to implement the different component, whole group versus small group.

This tool aims to help facilitate a conversation about each of the six elements as they apply to the essential learning standards for the unit. Your team's commitment to this discussion supports a commitment to ensuring meaningful, relevant, and equitable learning experiences for each and every student.

How you use these research-affirmed elements of effective lesson planning for improving student learning and deepening student perseverance is the next step to great lesson success. That is the professional work you will discover in part 2 of this book.

Directions: As a team, use this tool at the beginning of a unit. Brainstorm and discuss ideas you may have for each section. Remember, not all of the six lesson-design components need to be consistent among members of your team for each and every lesson; however, the more you share and discuss your individual ideas, the stronger the learning opportunities for all students will be.

Essential Learning Standard: List your first essential learning standard for the unit.

Academic Language Vocabulary: List the vocabulary specific to this essential learning standard.

Learning Target (List the learning target or targets under this essential learning standard in the order you intend to teach them.)	Number of Days of Instruction for Each Learning Target	Prior-Knowledge Warm-Up Activities for Each Learning Target	Cognitive-Demand Task Balance		Lesson Closure for Evidence of Student Learning
			Lower-Level-Cognitive-Demand Task Examples	Higher-Level-Cognitive-Demand Task Examples	

Essential Learning Standard: List your next essential learning standard for the unit.

Academic Language Vocabulary: List the vocabulary specific to this essential learning standard.

Learning Target (List the learning target or targets under this essential learning standard in the order you intend to teach them.)	Number of Days of Instruction for Each Learning Target	Prior-Knowledge Warm-Up Activities for Each Learning Target	Cognitive-Demand Task Balance		Lesson Closure for Evidence of Student Learning
			Lower-Level-Cognitive-Demand Task Examples	Higher-Level-Cognitive-Demand Task Examples	

Essential Learning Standard: List your next essential learning standard for the unit.

Academic Language Vocabulary: List the vocabulary specific to this essential learning standard.

Learning Target (List the learning target or targets under this essential learning standard in the order you intend to teach them.)	Number of Days of Instruction for Each Learning Target	Prior-Knowledge Warm-Up Activities for Each Learning Target	Cognitive-Demand Task Balance		Lesson Closure for Evidence of Student Learning
			Lower-Level-Cognitive-Demand Task Examples	Higher-Level-Cognitive-Demand Task Examples	

Essential Learning Standard: List your next essential learning standard for the unit.

Academic Language Vocabulary: List the vocabulary specific to this essential learning standard.

Learning Target (List the learning target or targets under this essential learning standard in the order you intend to teach them.)	Number of Days of Instruction for Each Learning Target	Prior-Knowledge Warm-Up Activities for Each Learning Target	Cognitive-Demand Task Balance		Lesson Closure for Evidence of Student Learning
			Lower-Level-Cognitive-Demand Task Examples	Higher-Level-Cognitive-Demand Task Examples	

Figure P1.3: Team unit-planning tool for overall lesson design.

Visit go.SolutionTree.com/MathematicsatWork for a free reproducible version of this figure.

PART 2

Team Action 4: Use Effective Lesson Design to Provide Formative Feedback and Student Perseverance

> The highest stake of all is our ability to help children realize their full potential.
> —Samuel J. Meisels

The Mathematics in a PLC at Work lesson-design tool discussed at the end of part 1 (figure 7.1, pages 52–53) supports the "What?" of lesson design. In part 1, you were asked questions about how you choose different elements of lesson design. You and your colleagues examined and reflected on six lesson-design elements for quality daily mathematics lessons.

1. Essential learning standards: the *why* of the lesson
2. Prior-knowledge warm-up activities
3. Academic language vocabulary as part of instruction
4. Lower- and higher-level-cognitive-demand mathematical task balance
5. Whole-group and small-group discourse balance
6. Lesson closure for evidence of learning

Now, in part 2, you and your colleagues will explore how you use your lesson-design choices for each lesson-design element during the lesson. You will encounter these team discussion tools along the way to support your team's formative assessment and intervention work.

- Checking for Understanding as Part of the In-Class Formative Assessment Process (figure P2.1, page 64)
- Prior-Knowledge Warm-Up Activity Process (figure 8.1, page 68)
- Vocabulary Instruction Activities and Graphic Organizers (figure 9.1, pages 73–75)
- Planning for Vocabulary Instruction (figure 9.2, page 76)
- Self-Evaluation Checklist on Whole-Group Questioning (figure 10.6, page 85)
- Whole-Group Discourse Teacher Prompts and Student Sentence Starters (figure 10.7, page 86)
- Assessing and Advancing Questions to Ask Students (figure 10.8, page 88)
- Walk-Around Formative Assessment (figure 10.9, page 89)
- Management of Student Teams (figure 10.13, page 93)
- Management of Student Teams (With Answers) (figure 10.14, pages 94–95)

- Sample Lesson-Closure Activities (figure 11.1, pages 100–101)
- Connecting In-Class Tier 1 Intervention to the End-of-Unit Assessment (figure 12.1, page 110)
- Using the Mathematics in a PLC at Work lesson-design elements (figure P2.2, pages 113–114)

Accomplished teachers of mathematics have a built-in intuition with both deep and surface understanding of the fundamental purpose of a mathematics lesson: *student demonstrations of learning through productive perseverance, corrective feedback, and formative refinement of their work.*

Thus, part 2 reveals a fourth team action for your work in mathematics: using the six lesson-design criteria to provide formative feedback and foster student perseverance during the lesson. Team action 4 ensures your collaborative team reaches clarity on how to effectively respond *in class* to the third and fourth critical questions of a PLC (DuFour et al., 2016).

3. How will we respond *in class* when some students do not learn?
4. How will we extend the learning *in class* for students who are already proficient?

How will you engage your students in the essential learning standard, and how will you collect evidence of learning during the lesson? How will you vary the types of activities and tasks to ensure you maintain student engagement and promote perseverance throughout the lesson? And, most importantly, how will you fully engage the students during the discourse of the lesson?

Part 2 provides you with discussion tools, samples, and insights that will help you to answer these questions for every mathematics lesson, every day.

Before you dive into the details of how, take a moment to examine your current *use* of the six mathematics lesson-design elements listed in the teacher reflection.

Maintaining student perseverance during a mathematics lesson can be a very difficult task. There is quite a bit of evidence that *formative feedback with student action* during the lesson can help you to overcome this daily student challenge in mathematics. Examining this evidence in part 2 begins by understanding the nature of FAST formative feedback as part of the lesson-design *process*.

> **TEACHER** *Reflection*
>
> How often do you currently implement the six lesson-design elements? Rank yourself and your colleagues on a scale of 1 to 10 for each criterion (1 = not at all; 5 = sometimes, and it could use improvement; and 10 = we are awesome at it and use it all the time).
>
> 1. Essential learning standards: The *why* of the lesson _____
> 2. Prior-knowledge warm-up activities _____
> 3. Academic language vocabulary as part of instruction _____
> 4. Lower- and higher-level-cognitive-demand mathematical task balance _____
> 5. Whole-group and small-group discourse balance _____
> 6. Lesson closure for evidence of learning _____

Essential Characteristics of Meaningful FAST Feedback

Team action 4, *use effective lesson design to provide formative feedback and student perseverance*, is all about moving learning forward. You and your team can provide deep support for student perseverance and proficiency with the use of effective formative assessment processes in the classroom. Typically, you might not think of feedback and assessment as a support for learning *during* instruction, but effective assessment (particularly formative assessment processes as part of instruction) is critical to the learning process (Kanold & Larson, 2012; NCTM, 2014). According to Dylan Wiliam (2011):

> When formative assessment practices are integrated into the minute-to-minute and day-by-day classroom activities of teachers, substantial increases in student achievement—of the order of a 70 to 80 percent increase in the speed of learning—are possible. . . . Moreover, these changes are not expensive to produce. . . . The currently available evidence suggests that there is nothing else remotely affordable that is likely to have such a large effect. (pp. 160–161)

The formative feedback and assessment process is much more than just observing evidence of student learning (checking for understanding), which is at best diagnostic. For the process to also be formative, you or your students' peers (or both) must provide meaningful feedback to each other during engagement with the mathematics tasks, *and* subsequently *take action* on that feedback.

Take a moment to reflect on the role of formative feedback in your classroom currently.

TEACHER *Reflection*

Think about when you introduce a mathematics task during a lesson. How do you know when students get stuck, and what happens if they do get stuck?

Describe how you currently provide feedback to students while they are working on the mathematics problems you have selected for the lesson.

_____
_____
_____
_____

Douglas Reeves (2011, 2016) and John Hattie (2009, 2012) provide insight into four essential characteristics of meaningful feedback that create a basis for effective formative feedback to students. Think of this as your attempt to provide FAST (fair, accurate, specific, and timely) corrective feedback to your students *during the lesson*. The four essential characteristics of feedback include:

1. **Fair**—Does your feedback rest solely on the quality of the students' work and not on other student characteristics? This includes some form of student comparison to others in the class.

2. **Accurate**—Is the feedback during the in-class activity actually correct? Do students receive prompts, solution pathway suggestions, and discourse that are effective for understanding the mathematical task or activity as you tour and check for understanding?

3. **Specific**—Does the verbal feedback students receive contain enough specificity to help them persevere and stay engaged in the mathematical task or activity process? Does the feedback help them to get "unstuck"—to advance their thinking as needed? (For example, "Work harder on the problem" is not helpful feedback for a student.)

4. **Timely**—As you tour the room and listen in on peer-to-peer conversations, is the teacher feedback immediate and corrective to keep students on track for finding a solution pathway?

These four characteristics describe how to give feedback; however, to have a magnified impact on student learning, students must also take action on the feedback. If during your best teacher-designed moments of classroom formative assessment processes, students fail to take action on evidence of their continued areas of difficulty, the learning cycle stops for the student.

During the lesson, imagine taking a close look at what your students are doing as the mathematics lesson progresses. Are your students being expected to embrace their errors as they *reflect, refine,* and *act* on the in-class practice with their peers and with you?

According to W. James Popham (2011), the following occurs when during the lesson teachers use formative assessment and feedback well:

> It can essentially *double the speed of student learning* [emphasis added] producing large gains in students' achievement; and at the same time, it is sufficiently robust so different teachers can use it in diverse ways and still get great results with their students. (p. 36)

So, what does this have to do with the work of your collaborative team? To start, your team should be clear about the difference between *checking for understanding* and formative assessment as part of instruction.

Checking for Understanding and the Formative Feedback Process

What is the difference between *checking for understanding* and *formative assessment*? Take a few moments to consider some of your current methods of checking for understanding in the teacher reflection (page 65).

The activity described in figure P2.1 (page 64) is designed to facilitate your team discussion about the differences between checking for understanding as a diagnostic tool and formative assessment processes as a student learning tool during mathematics instruction.

Directions: As a team, use this tool to engage in a conversation about the differences between checking for understanding and formative assessment processes.

1. How do you currently check for student understanding on a daily basis? Write your responses in the left column of the following chart. For each strategy, indicate whether you use:
   - **W** for whole-group discourse (teacher at the front of the room)
   - **S** for small-group discourse (students working with their peers)
   - **B** for both
2. How do you and your colleagues describe the difference between checking for understanding and formative assessment processes in the classroom during instruction?
3. How can you implement formative assessment processes in the classroom each day? Remember for the process to be formative, students must take action on the FAST feedback they receive.

In the right column, explain ways that you could extend each check for understanding to become a more formative learning moment with action by students.

Checking for Understanding	Formative Assessment Process
Example: Using whiteboards or an electronic smartboard, students work in teams to find a solution pathway to the math problem presented and display solutions upon command from the teacher.	**Example:** Without judgment on the correctness of the solution, use the whiteboard responses to regroup students and ask them to each defend their solution, with a chance to correct potential errors and rewrite the solution based on feedback from peers.

Figure P2.1: Team discussion tool—Checking for understanding as part of the in-class formative assessment process.

*Visit **go.SolutionTree.com/MathematicsatWork** for a free reproducible version of this figure.*

As you complete the questions and fill in your responses to the chart, consider how you can extend your various checks for understanding to become part of a more formative and FAST feedback process.

As teachers of mathematics, you and your colleagues need to continue your various checks for understanding. However, realize it is only the starting point of the learning process for students during the lesson. Checking for student understanding at most has a diagnostic impact and helps you to modify instruction as needed. Consider this story from author Tim Kanold. He provides an analogy that differentiates between checking for understanding and formative assessment processes as part of the mathematics lesson design.

Thus, for your students there is minimal impact on learning unless you take the check for understanding and extend that check to include a formative process with FAST feedback and action. This process, in turn, allows your students to embrace their mistakes (what will be their response during the mathematics lesson when they are not learning) and try again on that specific mathematics task during the lesson.

In-class formative assessment *processes* with student action on feedback support student perseverance on a task. As you and your team work together to implement more rigorous mathematics tasks that require students to use multiple strategies or representations and justify

TEACHER *Reflection*

How do you currently check for understanding during a lesson? List some of the current methods you use, such as whiteboards, thumbs up or thumbs down, clickers, or other electronic tools.

_____
_____
_____
_____
_____
_____
_____
_____
_____
_____

Personal Story TIMOTHY KANOLD

Imagine I went to the doctor for a "check for understanding" of my health. The doctor does a blood test (a check for understanding) and reveals to me that I have high blood pressure. She informs me that my heart health is not quite right, and then sends me on my way. In some ways it would be like you or I checking for understanding by asking students to show their responses or answers to a practice mathematics problem in class, and when some of the student responses are not quite right and filled with errors or incorrect solution pathways, we just move on to the next mathematics problem in the lesson.

My doctor's check for understanding was helpful for her responsibility as a doctor; she made a diagnostic observation of my health. However, for me, there is no learning or improvement at all unless my doctor also gives me FAST feedback (on the spot while I am still in her office), tells me three or four actions I need to take to correct my heart health, and I walk out and actually take action on her feedback. Without any action on my part, I will still have my high blood pressure problems.

The same is true for our students. As we check for their understanding during the lesson, we must also provide feedback on errors, and then expect students to own those errors and take action on improving their performance on the mathematics task at hand. Otherwise they will still have learning problems for that mathematics standard. Only when students act on our feedback does our lesson "form" their thinking or become formative.

their thinking, student perseverance via meaningful and FAST feedback becomes essential to the learning process.

As you work your way through suggestions for how to use each of the six elements of a highly effective mathematics lesson design, keep in mind the importance of each element and how it supports the formative and corrective feedback process—each and every day.

TEAM RECOMMENDATION

Use FAST Feedback During the Lesson

- Think about the type of feedback you provide students. Remember it must be fair, accurate, specific, and timely. Is there one area you need to improve upon?

- Decide how you can shift your current use of checking for understanding to become more formative for students.

CHAPTER 8

Essential Learning Standards and Prior-Knowledge Warm-Up Activities

I never teach my pupils. I only attempt to provide the conditions in which they can learn.

—*Albert Einstein*

The Albert Einstein quote in the epigraph describes a certain vision for mathematics instruction. What is your immediate reaction to his words? Do you agree? Do you disagree?

As you recall from part 1, chapters 1 and 2, the daily learning target and the prior-knowledge warm-up activities provide initial conditions and a context for learning the daily lesson. The mathematical tasks or prompts you choose as a warm-up activity need to connect the students to prior-knowledge standards and create a context for successfully entering into the learning targets for the lesson.

What types of mathematical tasks or problems will help *prepare* students to engage in the lesson in a meaningful way each day? To begin, as a team, use figure 8.1 (page 68) to discuss how you implement your lesson warm-up to support student engagement while also helping students connect to the learning target for the day's lesson.

TEACHER *Reflection*

What are some of your best ideas for creating a context for the essential learning standard for the lesson and helping your students understand why the learning target is necessary for the lesson?

How do you currently help your students to recall prior knowledge and connect that knowledge to the learning target for the lesson?

Directions: As a teacher team, discuss your current process for implementing a prior-knowledge warm-up activity using the questions that follow.

1. How do you currently choose your prior-knowledge warm-up task or tasks? Share your ideas as a team.

2. How do you structure the student discourse with peers during the warm-up (since it is a review activity)?

3. In what ways do you use the warm-up activity to assess student readiness for the lesson?

4. How do students know if their responses to the warm-up activity are correct?

5. What are you doing while the students are completing the warm-up (taking attendance, checking off homework, moving about the room checking in with student teams, and so on)?

6. If you see students struggling with the warm-up activity, how do you respond? How do they respond?

7. How do you currently use the warm-up activities to activate student discussion and create a context for the lesson standard of the day?

Figure 8.1: Team discussion tool—Prior-knowledge warm-up activity process.

Visit go.SolutionTree.com/MathematicsatWork for a free reproducible version of this figure.

Guidelines to Consider

As your collaborative team reflects on your responses in figure 8.1, there are four helpful guidelines to consider for your prior-knowledge and essential learning standard routines.

1. **Always use a small-group focus:** Small groups provide students with an opportunity to work together and practice communicating their ideas to review a concept from the prior lesson, unit, or year. Small groups also create a natural opportunity for students to provide each other with feedback and re-engage by reviewing content before exploring new content for the lesson. You should tour the students' peer-to-peer discussions to *see and hear* student understanding and provide small-group discourse feedback as students need it.

2. **Provide higher-level-cognitive-demand tasks and prompts:** The mathematical tasks or discussion prompts you choose for the warm-up activities should generally be rigorous and promote mathematical thinking and student understanding on previously learned standards. Warm-up activities should require your students to reason, justify, or problem solve—tasks that go beyond demonstrating a routine skill.

   The purpose of the prior-knowledge activity is to promote connections to the essential learning standard or new knowledge to be taught and the critical thinking to come ahead in the lesson. Try to avoid simple rote memorization of routine tasks; rather, present a question or task that seeks to determine what students *understand*.

3. **Structure a clear routine for the start of class:** There should be evidence the warm-up activity is built around a carefully selected mathematical task or prompt with well-organized and understood routines for how your students are to proceed, engage, and interact with each other.

   The warm-up activity should be readily available to students as class begins, with clear directions and prompts for how to proceed and share their thinking with one another. You should use no more than five to ten minutes of the overall lesson time as students respond to the discussion prompt or the mathematics problem you provided for connecting their prior knowledge and understanding.

4. **Do not "go over" the warm-up activity in class:** As you walk around the room and observe student teams successfully engaging in the warm-up activity, allow them to discuss the mathematics task or prompt with their peers and with feedback from you as you determine their readiness for the lesson. Do not "go over" the warm-up in class. Students can review answers you supply to check their work themselves. Or, as you walk around the room monitoring students, you may want to reveal a few student solutions to share on the document camera or other public display for students to review while they are discussing the prompts from the activity.

If you observe students struggling during the warm-up activity, this is a great opportunity to reassess your next steps to start the lesson. If there are just a few students struggling, then you may strategically pull a small group of students aside during an appropriate time during the lesson. If you see the majority of the class struggling, then stop the activity and ask questions, provide insight, or give students an additional scaffolding prompt to help them re-engage in the mathematical task.

For sample warm-up problems and prompts, see chapter 2, figures 2.1 through 2.4 (pages 21–23) in part 1. You and your colleagues should also use figure 2.5, the Prior-Knowledge Task-Planning Tool (page 25), as you brainstorm efficient and appropriate activities to connect student prior mathematical knowledge to the new and expected learning for the day.

Remember that your students come to the mathematics lesson with a broad range of pre-existing knowledge and skills. How well they persevere through, process, and integrate the new information from your daily lesson is influenced by the connections you help them make to previous learning during the warm-up activity.

Since the warm-up activity is designed to assess a prior-knowledge mathematics concept or skill, it is a natural outcome of the activity to reveal the new essential standard and learning target for the lesson that day once the warm-up activity is completed. It creates the *why* for the lesson and the learning progression context for the students.

For example, you could ask your students a question that sets up the context for the progression of the next learning standard connected to the day's lesson: "What do you remember about base ten from second grade? Well, today in this third-grade lesson on base-ten numbers we are going to study. . . ." Or, "Recall our discussion about equivalent expressions from the previous unit. What is meant by the word *equivalence*? In this lesson today we are going to study. . . ." Or, "Do you remember that linear functions represent a constant rate of change? Well, in today's lesson, we are going to study. . ."

In general, moving out of the prior-knowledge warm-up activity and into a public declaration for the purpose and context for this day's lesson—to learn the essential learning standard of the unit and the learning target for the day—is the best process to follow. Remember that to answer PLC critical questions 3 and 4—How will we respond in class when some students do not learn, and how will we extend the learning in class for students who are already proficient?—begins with PLC critical question 1, What is it you want students to know and be able to do, *today*?

However, sometimes the lesson may be best served if you delay *when* to reveal the essential standard to the students. It is possible that due to the exploratory nature of the lesson design, you might want to wait. Regardless of when you reveal the learning target, be sure to set the context for how the learning target "fits" into the student progression for learning (what we have learned previously, what we will be learning later on, and how the standard for this lesson fits *now*—today).

TEAM RECOMMENDATION

Prior-Knowledge Warm-Up Activities and the Essential Learning Standards

- As a team, understand how prior-knowledge warm-up activities provide the context for why the essential standard for the days' lesson is relevant and why students need to learn it.
- Prior-knowledge warm-up activities create an entry point into your lesson by activating students' knowledge of connecting concepts.
- The warm-up provides you an opportunity to assess student readiness for the lesson.
- The warm-up time is a great way to support student engagement in a collaborative process and develop student confidence.

Once you and your team discuss how you implement prior-knowledge warm-up activities (it can be very robust and varied), it is also important to discuss the strategies you use to support vocabulary and language development during the lesson for each and every student. The next chapter explores this idea in greater detail.

CHAPTER 9

Using Vocabulary as Part of Instruction

> Language is a major medium of teaching and learning mathematics; we serve students well when we support them in learning mathematical language with meaning and fluency.
>
> —Rheta N. Rubenstein

As discussed in chapter 3 (see part 1, page 27), it is important not to overlook planning for the vocabulary instruction necessary for students to learn a mathematics standard. Teachers have a tendency to plan for the mathematical tasks to teach the standard, but often overlook how to engage students in the vocabulary for the lesson. For this reason, this chapter provides a variety of vocabulary strategies for you and your team to consider.

To begin, take time to reflect on your current use of student vocabulary development within your mathematics instruction.

The verbal language of mathematics is significant for at least two reasons (Ben-Hur, 2006):

> First, it is significant because most of the verbal terms—such as *angle, area, slope, function, product, quotient, mode, factor, variable, tangent, polygon, prime number, permutation, cosine, irrational number,* and many others—function as "conceptual packages." Terms such as these do not represent particular objects. They are abstract categories that conceptually organize mathematics. Second, the verbal language is significant to mathematics simply because that language is the essential tool of dialogue about specific mathematical ideas—a tool of teaching and learning. (pp. 66–67)

Direct vocabulary instruction can improve mathematical understanding as well. Irene Miura and Jennifer Yamagishi (2002) find that direct instruction in language, symbols, and their connections can help your students acquire concepts.

It can take up to twelve experiences for a word to have meaning for a student (McEwan-Adkins, 2010). This is why tools like graphic organizers and active word walls where students receive time for their own sense-making and reasoning about the words or symbols are an important tool for continued use of vocabulary words and mathematics notation in classroom discussions and lessons.

TEACHER *Reflection*

How do you and your collaborative team members currently use vocabulary instruction as a way to maintain student perseverance and engagement throughout a mathematics lesson and during the unit of instruction?

_____

_____

_____

_____

_____

_____

> ## Personal Story JESSICA KANOLD-McINTYRE
>
> As a middle school mathematics teacher, I remember learning that vocabulary instruction was important. I would start a word wall with great intention and then by the second or third unit of the year, I would forget about updating the words.
>
> I tried to embed vocabulary instruction into my day-to-day lessons, but I did not realize the importance of *frequency*. I would embed some vocabulary into my lesson, provide an example or a definition, and expect students to understand and reference the word on their own.
>
> As I have learned more about working with the language and reading needs of my students, I have discovered they need time to make their own meaning with new vocabulary while also being encouraged to use the words throughout the instructional process. This means that just having students use the glossary in the book to write the definitions of words or complete a crossword puzzle with vocabulary words is not enough.

In order to support equitable student experiences with language between teachers, it is important for your collaborative team to create common expectations around vocabulary instruction and identify the new vocabulary on a unit-by-unit basis. As part of the unit-planning process at the beginning of a unit, you and your team should identify and discuss vocabulary terms to ensure consistency around the use of terms for all students in the grade level or course.

To identify vocabulary terms for a unit, it is best to go back to the actual formal essential unit standards for direction. You should have written the vocabulary for the unit on your unit plan document. Then, as you create each lesson, note which vocabulary words you plan to highlight within the lesson, especially relating the vocabulary words to the possible challenges students may have with that particular word.

Take a moment to consider how you choose vocabulary for your instruction in the teacher reflection.

When and How to Teach Vocabulary

You may be wondering, "When is the best time during a mathematics lesson to tackle vocabulary instruction?" It depends. In *Visible Learning for Mathematics*, John Hattie, Douglas Fisher, and Nancy Frey (2017) report that you can use vocabulary instruction at almost any time during a lesson. First, you can use it as a preteaching activity as the lesson begins and based on possible feedback you received during the prior-knowledge warm-up activity. Second, you can use it to reinforce just-in-time learning during the lesson as you need it. Third, you can reinforce vocabulary by formalizing meaning of key words at the end of a lesson as part of a closure activity.

There are several activities or strategies you can use to inform your thinking about vocabulary instruction. Table 9.1 identifies various types of vocabulary

TEACHER *Reflection*

How do you currently choose vocabulary for your units or lessons? Are the words you choose consistent with those of other teachers across your grade level or course?

Table 9.1: Vocabulary Challenges and Focused Strategies

Challenge	Focused Strategy
Some words are shared with everyday English, science, or other disciplines and may have distinct or more technical meanings within mathematics.	Be aware of potential confusion. Distinguish the technical from the everyday meanings. Help students understand what the terms share and the reason why the common language term was adopted for mathematics.
Some words are found only in mathematics.	Help students see the roots and origins of mathematics terms. Point out common English words with the same root. Help students see how the roots build the mathematical meaning.
Some words have more than one mathematical meaning.	Remind students of the multiple usages of words and to use context clues to know which is the intended meaning. Help students see why the word makes sense in each context.
Some words are learned in pairs that often confuse students.	If possible, separate the learning of the two terms so that students understand one word well before introduction of the second word. Continue to use word origins and relate each of the two terms to everyday English words. Acknowledge the challenge, and have students double check one another when they use the words.
Some words sound like others (homonyms and near homonyms).	Say the words clearly. Spell them. Distinguish them. Use each in its particular context. Draw a picture.
Sometimes modifiers change meanings of words in critical ways.	Have students explore the unmodified term, the modifier, and then the full phrase (for example, *bisector*, *perpendicular*, and then *perpendicular bisector*) to help them see the broad category as well as the specific meaning within it.

*Visit **go.SolutionTree.com/MathematicsatWork** for a free reproducible version of this table.*

challenges students may have and provides language strategies to address each issue as a way to support your instruction (Rubenstein, 2007). Examples of these vocabulary challenges appear in figure 3.1 on page 29.

In addition to the suggestions in table 9.1, figure 9.1 presents several activities and graphic organizers you can use to support and ensure student engagement in the vocabulary for the lesson.

Activities
Directions: Of all the vocabulary strategies listed, which types do you currently use during your mathematics lessons as part of instruction? Which might you try?
Cloze Passages Use cloze passages for steps to build student familiarity with text structures, vocabulary, and comprehension. 1. Retype a passage and place a blank line in place of strategic words. 2. Have students read the passage and try to determine from context the words that might fit in the blanks. Students earn five points for using words from the textbook and three points for using another word that makes sense to them in the context of the passage. 3. Read the correct passage to the whole class and allow students to score their work. Discuss whether words students chose that differ from the author's make sense in the context of the passage. 4. Have students total their points to determine the "winner."
Survival Words This six-step activity uses inquiry to determine how well students may understand vocabulary words they need to successfully read a mathematical task. 1. Choose several words that may trip students up and that students are likely to see again, or use a list of high-frequency words from the mathematics unit the class generates. 2. Have students make a chart that has the following six column headings: Word, A, B, C, D, and Meaning.

Figure 9.1: Team discussion tool—Vocabulary instruction activities and graphic organizers.

continued →

3. Have students copy each word down in the first column of the chart and check the appropriate A, B, C, or D category for each word.
    A. I know the meaning, and I use the word.
    B. I know the meaning, but I don't use the word.
    C. I've seen the word before, but I don't really know it.
    D. I've never seen the word before.
4. Ask students to write the meanings of as many of the words as they know in the meaning column.
5. Break students into groups to share the words and meanings they are most confident about.
6. Go over their charts with students and answer any questions, giving additional information or clarification as they need it.

Making Meaning

This five-step activity is especially effective with English learners, but it also helps any student to make meaning of vocabulary.
1. Divide a 3 × 5 note card into quadrants.
2. In the upper-left quadrant, ask students to write the vocabulary word in English or his or her most comfortable language.
3. In the upper-right quadrant, ask the student to write a synonym for the vocabulary word or to write the word in his or her most comfortable language.
4. In the lower-left quadrant, ask the student to define the word in any language using his or her own words, not a verbatim definition from a text.
5. In the lower-right quadrant, ask the student to make meaning of the word with a picture or example.

Word Wall

In this activity, the word wall displays words used throughout a unit. Students give meaning to the words on display.
1. Identify common unit vocabulary words.
2. List the words on chart paper or on a whiteboard in the classroom. Keep the word wall in the same part of the classroom all year.
3. As words from the unit appear during instruction, students write the meanings of the words and write an example or draw a picture in their notes or math journal to make sense of the word. One student documents his or her work on the class chart paper.

Tip: Reference the word wall routinely in class.

At the beginning of the unit, use a graphic organizer like a KIP chart that follows to help students make meaning of the words.

At the middle and end of the unit, use a graphic organizer like a foldable (see the example that follows) to help students connect the words and show relationships.

Graphic Organizers

KIP

KIP is a graphic organizer that uses a chart to document important vocabulary: key vocabulary, information about the vocabulary, and a picture drawing of the word. The five steps are:
1. Create a chart like the one that follows.

Key	Information	Picture
Example:		

2. Students document the key vocabulary (K) in the chart. Each word will have its own chart.
3. Students write information (I) about the word—their own definition for the word.
4. Students draw a picture (P), if appropriate, for the word.
5. Students write or draw an example showing the word.

Highlights

This four-step graphic organizer encourages students to read problems thoughtfully and document what the problem is asking.
1. Have students read directions to a problem set or a word problem.
2. Have students underline the important information and provide a clarifying symbol if possible (for example, put a + above the word *sum* or = above the word *is*).
3. Have students circle what the problem requires them to solve for.
4. Discuss how problems are similar or different and help students use the vocabulary to understand the type of problem that is being asked.

Frayer Model
This is a graphic organizer for vocabulary.
  1. Have students write the word they are going to define in the middle of the graphic organizer.
  2. In the upper left corner, have students write the definition of the word (in their own words).
  3. In the upper right corner, have students write the facts or characteristics they know about the word.
  4. In the lower left corner, have students write or draw an example of the word.
  5. In the lower right corner, have students write or draw a nonexample of the word.

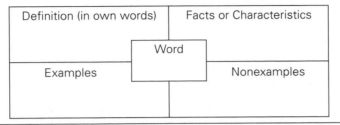

Foldable
A foldable is a tool used to organize vocabulary for any unit. Most foldables require one sheet of paper that you fold or cut to creatively organize words, definitions, and examples. You can use foldables to integrate reading, writing, thinking, organizing data, researching, and other communication skills into an interdisciplinary mathematics curriculum (Zike, 2003). Visit www.pinterest.com/k8ekakes/math-foldables for more information about using a foldable.

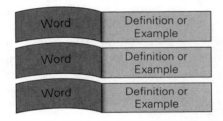

Source: Dale & O'Rourke, 1986; Frayer, Fredrick, & Klausmeier, 1969.

Visit **go.SolutionTree.com/MathematicsatWork** *for a free reproducible version of this figure.*

These activities and graphic organizers provide a sampling of ideas you could use to embed vocabulary instruction purposefully into your lesson plans.

Figure 9.2 (page 76) applies the activities and organizers from figure 9.1 to sample formal unit standards from grade 4, middle school, and high school to show what strategies a team might choose for a specific unit standard. The tool in figure 9.2 is one that you and your team can use as you plan for vocabulary instruction within your units.

In order to use figure 9.2 as a team-planning tool, you could list all the formal unit standards for one unit in the left column, and then specifically note the vocabulary for the unit in the middle column and how you plan to engage students in the vocabulary in the right column.

The goal of vocabulary instruction is to provide opportunities for students to create their own meaning, so they can confidently use the words they learn in written and oral communication. As you embed vocabulary instruction into your lessons, you will find that you might want to formally assess vocabulary words on your unit assessments, but it is not essential to do so. However, students need to be able to read the words and understand them, hear the words and understand them, and use the words in explanations and work. Vocabulary is crucial to students being able to listen, talk, and learn from one another through discourse and peer review.

Recall that part of the purpose of your work with implementing these lesson-design criteria is answering the third critical question of a PLC, What will be my in-class response when you do not know what I need you to learn? Sometimes, that response needs to be to develop your academic vocabulary and allow you time to process understanding on mathematics notations being used and what they mean. Terms like *sum* and *denominator*, *equivalence* and *greater than*, *rational number* and *translation*, or *slope field* and *derivative* can cause confusion for students as a special part of the mathematics language they learn.

Formal Unit Standard	Vocabulary	Vocabulary Strategy
Grade 4 Explain why a fraction $\frac{a}{b}$ is equivalent to a fraction $\frac{(n \times a)}{(n \times b)}$ by using visual fraction models, with attention to how the number and size of the parts differ even though the two fractions themselves are the same size. Use this principle to recognize and generate equivalent fractions.	Fraction Equivalent fraction Fraction model Numerator Denominator	A graphic organizer with visuals would be extremely helpful for students in the fraction unit. The KIP model, Frayer model, or a foldable would all be appropriate strategies to help students organize the words and give them time to make their own meaning. The term *fraction* was previously introduced in grade 3; however, it is important that students understand not just the word, but also how to represent a fraction visually and be able to provide multiple examples.
Middle School Understand the concept of a unit rate $\frac{a}{b}$ associated with a ratio $a : b$ with $b \neq 0$, and use rate language in the context of a ratio relationship.	Ratio Rate Unit rate	These words are new for students in sixth grade, so it is important to spend time guiding them through different activities to make meaning of the words in isolation and within the context of real-world problems. The KIP model or Frayer model would be appropriate strategies to help students organize the words and give them time to make their own meaning. A word wall would also help keep the terms visible for students throughout the unit and for the whole year.
High School Solve linear equations and inequalities in one variable, including equations with coefficients represented by letters.	Coefficient Terms Factors Variable Expression Linear Equation Inequalities	Even though some of these words aren't new for students in high school, a foldable would be an appropriate graphic organizer to help make sense of the language. In addition, students could use it as a tool to reference when working on tasks. The words *expression*, *variable*, and *terms* might have different meanings for students outside of mathematics, so it is important to point out the differences.

Figure 9.2: Team discussion tool—Planning for vocabulary instruction.

Visit **go.SolutionTree.com/MathematicsatWork** *for a free reproducible version of this figure.*

Accomplished teachers of mathematics place a lot of energy into using and managing small-group discourse and providing opportunities for meaningful formative feedback to students during both the lower-level- and higher-level-cognitive-demand tasks chosen for each lesson. There are different demands on your lesson design for whole-group discourse activities versus small-group discourse activities. That discussion is next.

TEAM RECOMMENDATION

Use Vocabulary as a Part of Instruction

Consider the following recommendations with your team.

- As a grade-level or course-based team, select the vocabulary words in advance of the unit and post or give them to students on the day the unit starts.
- As you teach the unit, be sure to refer to the words, adding definitions, giving context, and allowing time for students to make meaning.
- Through the use of the mathematical tasks you choose and small-group discourse, assess for student understanding of the vocabulary, while also monitoring student misconceptions.
- As you teach, use graphic organizers such as the ones figure 9.2 describes to help clarify the meaning of words and support long-term retention of vocabulary.

CHAPTER 10

Implementing Mathematical Task and Discourse Balance

> Effective teaching of mathematics engages students in solving and discussing tasks that promote mathematical reasoning and problem solving and allow [for] multiple entry points and varied solution strategies.
>
> —National Council of Teachers of Mathematics

In chapter 4 (page 33) in part 1 of this book, you and your team discussed the nature of mathematical tasks, and the difference between lower-level-cognitive-demand tasks and higher-level-cognitive-demand tasks. Now, you will have a chance to examine how to implement those tasks in a way that maintains the rigor you and your team intended when you designed your plan for using the tasks to develop student learning of the standard.

Part of the struggle with implementing higher-level-cognitive-demand tasks is maintaining student perseverance throughout the task. Have you ever had students shut down when you try something new in class or when you give them a problem they perceive as too hard for them? If you have, then you know it is not fun when it happens. It's easy to think the problem was too hard and your students just can't do "those types of math problems."

In order to support student perseverance through the higher-level-cognitive-demand mathematical tasks, you and your team members teach students how to act and respond like mathematicians. Does that surprise you?

Mathematicians are able to communicate their reasoning using accurate vocabulary. Mathematicians understand and see the importance in the process of learning and are able to reason through their own ideas and learn from early mistakes or errors while listening to and learning from others.

Mathematicians are able to use and connect multiple representations and solve problems in many ways. When students learn to represent, discuss, and make connections among mathematical ideas in multiple forms, they demonstrate deeper mathematical understanding and enhanced problem-solving abilities (Donovan & Bransford, 2005; Janvier, 1987).

To facilitate student perseverance during the lesson, there are five practices you can use to maintain student engagement and effort.

1. Connect the mathematical task to the *essential learning standard* for the unit and the daily learning target for the lesson.

2. Present mathematical tasks during the lesson that are of *higher-level cognitive demand*.

3. *Monitor* students' *initial* responses to the mathematical task through observation.

4. Ensure the mathematical task expects a process of student *reflection* and *communication with peers*.

5. Expect students to *compare and contrast solution pathways* to the mathematical tasks.

In order to engage in these five perseverance practices, it is necessary to be clear on the essential learning standard or standards for the lesson and be sure you and your team choose tasks that will support student engagement on that standard.

Thus, every day, you make a clear communication of the standard (an *I can* statement) to the students and connect it directly to the mathematical tasks you choose for the lesson.

Consider the five higher-level-cognitive-demand tasks in figure 10.1, figure 10.2, figure 10.3, figure 10.4 (pages 80–81), and figure 10.5 (page 81), representing kindergarten; grades 2, 4, and 7; and high school. A critical first step in selecting and planning to use higher-level-cognitive-demand tasks in class occurs when your team works through the solution pathways for the mathematical task together and discusses the nuances of the problem. You do this team activity *before* you use the mathematical task with your students.

Your collaborative team can use the sample tasks from figures 10.1–10.5 to explore elements of planning for the formative assessment process. As you use higher-level-cognitive-demand mathematical tasks within the context of what students are to *say and do*, your plans for creating a formative assessment process in class can unfold. Visit **go.SolutionTree.com /MathematicsatWork** for additional grade-level sample tasks similar to figures 10.1–10.5.

Kindergarten	
Standard: Represent addition and subtraction with objects, fingers, mental images, drawings, sounds (e.g., claps), acting out situations, expressions, or equations.	
Mathematical Task: Read the problem to the student: Julia has 9 cupcakes. She shares 4 cupcakes with her friends. How many cupcakes does Julia have now? Show your thinking with objects, words, pictures, or numbers.	
What types of misconceptions do you anticipate students will struggle with during the task?	
What types of scaffolding questions can you ask students to help guide their work on this task?	
How do you plan to provide feedback to student solution pathways and explanations?	
How will students work on the task with their peers and receive feedback from one another?	
How will you ensure all students take action on feedback received during the task?	

Figure 10.1: Planning for the formative assessment process—kindergarten.

*Visit **go.SolutionTree.com/MathematicsatWork** for a free reproducible version of this figure.*

Implementing Mathematical Task and Discourse Balance | 79

Grade 2	
Standard: Explain why addition and subtraction strategies work, using place value and the properties of operations.	
Mathematical Task: Adam bought 17 tickets at the fair. His brother gave him 9 more. Chelsea had 9 tickets from last year, and she bought 17 tickets when she got to the fair. Do Adam and Chelsea have the same number of tickets? How do you know? Explain using pictures, numbers, and/or words.	
What types of misconceptions do you anticipate students will struggle with during the task?	
What types of scaffolding questions can you ask students to help guide their work on this task?	
How do you plan to provide feedback to student solution pathways and explanations?	
How will students work on the task with their peers and receive feedback from one another?	
How will you ensure all students take action on feedback received during the task?	

Figure 10.2: Planning for the formative assessment process—grade 2.

*Visit **go.SolutionTree.com/MathematicsatWork** for a free reproducible version of this figure.*

Grade 4	
Standard: Understand a fraction $\frac{a}{b}$ with $a > 1$ as a sum of fractions $\frac{1}{b}$.	
Mathematical Task: Decompose the fraction $\frac{9}{12}$ in two different ways. Show your work. Be sure to explain how you decomposed your fraction.	
What types of misconceptions do you anticipate students will struggle with during the task?	
What types of scaffolding questions can you ask students to help guide their work on this task?	

Figure 10.3: Planning for the formative assessment process—grade 4.

continued →

How do you plan to provide feedback to student solution pathways and explanations?	
How will students work on the task with their peers and receive feedback from one another?	
How will you ensure all students take action on feedback received during the task?	

Visit go.SolutionTree.com/MathematicsatWork for a free reproducible version of this figure.

Grade 7

Standard:
1. Compute unit rates associated with ratios of fractions, including ratios of lengths, areas, and other quantities measured in like or different units.
2. Recognize and represent proportional relationships between quantities.

Mathematical Task: YoYo Yogurt sells yogurt at a price based upon the total weight of the yogurt and the toppings in a dish. Each member of Gia's family weighed his or her dish and paid the corresponding amount shown in the chart below.

Cost ($)	5	4	2	3.20
Weight (oz)	12.5	10	5	8

Does everyone pay the same cost per ounce? Show your work and explain how you know.

Is the cost proportional to the weight? Explain your answer.

What types of misconceptions do you anticipate students will struggle with during the task?	
What types of scaffolding questions can you ask students to help guide their work on this task?	
How do you plan to provide feedback to student solution pathways and explanations?	

How will students work on the task with their peers and receive feedback from one another?	
How will you ensure all students take action on feedback received during the task?	

Figure 10.4: Planning for the formative assessment process—grade 7.

*Visit **go.SolutionTree.com/MathematicsatWork** for a free reproducible version of this figure.*

High School
Standard: Derive the equation of a circle of given center and radius using the Pythagorean theorem; complete the square to find the center and radius of a circle given by an equation.

Mathematical Task: $x^2 + y^2 - 6x + 8y = 144$ The equation of a circle in the *xy*-plane is shown above. What is the *diameter* of the circle? Be sure to show your work and explain your answer.	
What types of misconceptions do you anticipate students will struggle with during the task?	
What types of scaffolding questions can you ask students to help guide their work on this task?	
How do you plan to provide feedback to student solution pathways and explanations?	
How will students work on the task with their peers and receive feedback from one another?	
How will you ensure all students take action on feedback received during the task?	

Figure 10.5: Planning for the formative assessment process—high school geometry.

*Visit **go.SolutionTree.com/MathematicsatWork** for a free reproducible version of this figure.*

Personally working in advance to solve and understand each mathematical task you plan to use for the lesson provides insight into the extent to which the task will engage students in the intended mathematics concepts, skills, and mathematical thinking, and reveals to you how your students might struggle. Working the task with your team provides information about possible solution strategies or pathways students might demonstrate. You and your team can also share anticipated misconceptions to assist with developing questions and responses to address possible challenges.

Support Student Perseverance During Whole-Group Discourse

As you move into the heart of the lesson using the mathematical tasks you have chosen for the students, you should ask yourself, "How will I check for student understanding during the lesson?"

Checking for understanding using verbal cues such as, "Did I go too fast?" "Isn't this an easy one?" "Okay?" "Everyone see that?" "Who doesn't understand that?" or, the often included rhetorical "Any questions?" is not sufficient. These particular cues set up two counterproductive conditions in your classroom: they (1) may not be accurate as a check for understanding and (2) they fall far short of your instructional goal to create and use a more formative feedback process with your students.

Reflect for a few moments on your questioning style during whole-group instruction.

Although it is difficult to do with much accuracy, there are two effective ways to check for understanding with feedback from the front of the room: (1) reflective summaries during presentation of new content and (2) effective questioning.

Reflective Summaries During Presentation of New Content

During a complex higher-level-cognitive-demand task, you can help students set up the initial investigation of the problem through your directed model, and then allow students some small-group time to discuss strategies for solving the task as you monitor their initial reactions and thoughts. This a good time to circulate among students to find out if they understand the set-up of the problem and possess the skills necessary to work a strategy of attack. While walking around, you receive and give feedback to more accurately pace the lesson. Students can work alone initially as they process their thinking, then compare and contrast solutions with peers on your command to do so.

Personal Story **TIMOTHY KANOLD**

I remember 1980 like it was yesterday. I was in my seventh season of teaching and my first year at West Chicago High School District 94. I attended our annual Illinois Council of Teachers of Mathematics (ICTM) meeting and received a thirty-page pamphlet from NCTM, our national organization. It was called *An Agenda for Action* (NCTM, 1980). A primary recommendation was this: *"Teachers should provide ample opportunities for students to learn communication skills in mathematics. They should systematically guide students to read mathematics and talk about it with clarity"* (p. 12).

My initial response was, "I do not know how to do this. But I am for sure going to try," and my mathematics lesson process was forever changed. My first action was to eliminate student callouts in class and engage my students in more peer-to-peer discussions during my questioning. I would redirect a student response for other students to evaluate, and they were surprised that I expected them to actually listen to one another and engage. I moved my body out of the front of the room and used more of a Socratic style as I helped students to rethink their mistakes. The result?

Students started to own their learning and not just depend on me for every moment of the lesson. In the beginning, they thought I was working them too hard because I was requiring them to actually *engage* in the lesson. Presenting solutions publicly (such as thirty students at a time standing at boards around my classroom) became a way of life during my daily lessons.

> **TEACHER** *Reflection*
>
> In order to check for student understanding during the lesson, what types of whole-group questions do you find you ask the most?
>
> _____
> _____
> _____
> _____
> _____
> _____
> _____

Effective Questioning

Direct questioning, often done by you from the front of the classroom, is so important to effective instruction that it is one of eight research-affirmed instructional strategies that NCTM (2014) outlines in its influential publication *Principles to Actions*:

1. Establish mathematics goals to focus learning.
2. Implement tasks that promote reasoning and problem solving.
3. Use and connect mathematical representations.
4. Facilitate meaningful mathematical discourse.
5. Pose purposeful questions.
6. Build procedural fluency from conceptual understanding.
7. Support productive struggle in learning mathematics.
8. Elicit and use evidence of student thinking.

Questioning during the presentation of new content helps you assess student awareness, readiness, and understanding. An effective questioning cycle would expect all students to listen actively both to the question as well as *to other students' responses*, even though individual students are responding to specific questions. Can you use strategies that encourage *all* students to consider the questions you ask the whole class? Yes, but whole-group questioning can severely limit student engagement in class.

The questions you pose should require more than one-word answers. For example, the question "What is the side length?" when the picture students are referencing shows 5 centimeters as the side doesn't promote meaningful or rich discourse. If you expect meaningful discourse, then questions must authentically engage students in discussion. For example, a question like, "What problems have we solved previously that are similar to this one? How are they the same? How are they different?" creates an open opportunity for students to explore and discuss.

A questioning cycle likely to result in this active engagement during whole-group discourse includes the following four steps.

1. Pose the question.
2. Provide wait time after each question to prevent student callouts (three or more seconds).
3. Select students randomly, making certain to include all students. Call on volunteers (raised hands) as well as nonvolunteers. Who gets called on in class sends important messages about students' mathematical identities (Smith et al., 2017). "By carefully listening to and interpreting student thinking, teachers can position students' contributions as mathematically valuable and as contributing to the broader collective understanding of mathematical ideas" (Smith et al., 2017, p. 165).
4. Redirect the student response to other students for their judgment of correctness or for an extension of an answer. When you explicitly ask other students to comment on a student's work, you encourage a "diversity of views and strategies in the discussion" that, if effectively done, can "promote historically marginalized student populations or students who do not have a strong record of success in mathematics" (Smith et al., 2017, pp. 165–166).

During whole-group discourse, some students are reluctant to wait to be called on and like to call out an answer or response. Student callouts can be disruptive because these students control the pace and the direction of the classroom discussion, which then takes away from the learning for those students who need more time to process.

When wait time goes unmonitored by you, your students usually receive less than one second to respond to a question. The use of wait time, at least three to seven seconds and upward to ten seconds, accompanied by high-order questions, results in students responding with more thoughtful answers and an increase in achievement (Cawelti, 1995).

A technique that also encourages wait time is to immediately follow a question with a phrase such as, "Raise your hand when you're reasonably sure of the next step or response to my question." *Reasonably sure* indicates it is okay to take a risk and possibly be wrong. The following questioning sequence illustrates an effective question cycle. This particular cycle of questioning reduces student callouts as well.

Teacher: "Class, what would be an example of a triangle with an area of 12 cm^2? Please draw and label a diagram on your paper and raise your hand to respond." [*The teacher walks away from the front of the room and monitors students as hands are raised.*]

Teacher: "Only John and Mia have such a triangle? Are there more? Three hands, four hands, anyone else?" (The total wait time should be three to five seconds.)

Teacher: "Okay, Alex, I notice you chose a right triangle. Please explain your diagram to the class." (Alex does not have his hand up, but the teacher noticed his work and wants him to be able to share with the class.)

Alex: "I drew a right triangle with legs of length 8 cm and 3 cm."

Teacher: "How many agree with Alex's example? Raise your hands! Who disagrees with Alex's example?" (This keeps all students accountable to listening to Alex's response.)

Teacher: "Who can prove or disprove Alex's assertion . . . Karina?" [*Karina raises her hand.*]

Karina: "In a right triangle, the legs represent a base and a height. Thus, $A = \frac{1}{2} bh$ or $A = \frac{1}{2}(8)(3)$, which is 12cm^2."

Teacher: "Do you agree with Karina's explanation? Raise your hand if you agree! Great! We all agree. As I walked around, I noticed all of you used a right triangle. Can you think of an example that is not a right triangle?" [*The dialogue continues around this advancing question from the teacher.*]

This whole-group questioning cycle allows you to know which students are "with you" during the lesson. It also increases wait time (thus giving students more time to actually think about a response) by extending the conversation on a topic, which supports the learning for all students in class.

There are some students in class that may need up to fifteen seconds of wait time to respond. Thus, when you can use a question or multiple questions to *facilitate a peer-to-peer conversation* about the content rather than a correct answer followed by another question, you are able to allow more time for all students to make sense of the standard and the mathematics tasks you use in class.

Figure 10.6 provides a *self-evaluation checklist* on your whole-group questioning techniques. Individually, reflect on your current classroom discourse. Then, as a team, share areas of strength and areas where you can continue to support each other as you continue to work on questioning.

As you look at your responses to figure 10.6, if your answers to every question aren't a 3 (always), there's room for improvement! Whole-group questioning and facilitation require thoughtful planning and purposeful teacher response and guidance.

What types of questions are the best to use to help facilitate a whole-group discussion where more than one student response is heard? Are there strategies and sentence starters students can use to respond to each other?

In figure 10.7 (page 86), you will find some sample teacher prompts and student sentence starters. As you read through the two lists, circle one or two new ideas you could bring back to your classroom to try.

Directions: Read each reflection question related to how you use questions in lessons. Identify whether you never, sometimes, or always employ the questioning strategy.

Reflection Question	1 Never	2 Sometimes	3 Always
1. Do I avoid asking, "Do you have any questions?"			
2. Do my questions promote total student involvement?			
3. Do I allow students to use think-pair-share before responding to questions?			
4. Do I require my students to act maturely when a student gives a wrong answer?			
5. Do I redirect certain questions and responses back to the entire class?			
6. Do I create a classroom environment that makes it safe for students to be wrong and even celebrate errors as a learning opportunity?			
7. Do I pause or give at least three to five seconds of wait time before calling on a student?			
8. Do I allow students to complete their answer before I jump in?			
9. Do I allow students to respond to another student's responses before I make a comment myself?			
10. Does my questioning give me meaningful input about the students' understanding of the concepts I'm teaching?			
11. When there are only a few hands raised to respond to a question, do I provide alternative ways to respond in order to get more students to participate?			
12. When only one student can answer a question, do I use this input and help others to understand and become involved in the question?			
13. Do I allow students to discuss ideas with their partners before asking a particular student to share ideas with the entire class?			

Directions: Once you complete the checklist, note both your strengths and weaknesses in your current whole-group question routines.

Directions: Compare your responses with a trusted team member and ask him or her if he or she would observe your lesson and let you know the level of student engagement during your whole-group questioning routines.

Figure 10.6: Team discussion tool—Self-evaluation checklist on whole-group questioning.

Visit go.SolutionTree.com/MathematicsatWork for a free reproducible version of this figure.

> **Directions:** Which of the following prompts do you currently use for whole-group discourse? List other ideas you use that are not in the list.
>
> *Teacher* questions to facilitate a discussion in whole-group discourse:
> - Do you agree or disagree with _____ and why?
> - Who can add on to what _____ said?
> - What is another way to say what _____ said?
> - _____, can you repeat what _____ just said?
> - Can you say that in your words?
> - Will this strategy always work?
>
> *Student* sentence starters to support student mathematics talk in whole-group discourse:
> - I respectfully agree with _____ because _____.
> - I respectfully disagree with _____ because _____.
> - I understand what you're saying, but I have a different idea . . .
> - Wait, I'm not sure I understand. Can you repeat what you just said?
> - So, I hear you saying . . .
> - How did you get that?
>
> List other questioning prompts or student sentence starters that you use:

Figure 10.7: Team discussion tool—Whole-group discourse teacher prompts and student sentence starters.

*Visit **go.SolutionTree.com/MathematicsatWork** for a free reproducible version of this figure.*

For your feedback to be effective, even during whole-group discourse, your students must take action on the feedback they receive, either from you as the teacher or from their peers. This action is very difficult for you to elicit from the front of the room during whole-group discourse.

The next section brings into focus the only way to design your lesson so that you can see and *hear all* students in a more fully engaged classroom climate: small-group discourse.

Support Student Perseverance During Small-Group Discourse

You may be great at managing whole-group discourse. However, it is almost impossible to make it part of a *formative learning process*. You can do great checks for understanding, but the learning of a mathematics task needs to be formative for every student. It is best for students to experience the learning of mathematics as part of small-group discourse activity in every lesson.

This is not a new concept. As far back as 1978, small-group discourse has been an expectation of expert mathematics instruction. According to Lev Vygotsky (1978), people learn complex knowledge and skills through social interaction.

Promoting and using small-group discourse throughout your lessons encourages student perseverance while also maintaining rigor. It authentically promotes a community of learners who see each other as valuable resources and communicate about their ideas. It provides multiple opportunities for students to reason and make sense of the mathematics. However, to support discourse in your classroom, you must understand the demands students experience as sharers and listeners (Kazemi & Hintz, 2014).

Going beyond checking for understanding from the front of the classroom and moving into the more beneficial formative feedback process as part of instruction require intentional development of rich mathematical tasks that align to the essential learning standard and are supported through robust student discourse. This means supporting student-engaged explorations and discussions with peers.

Use the teacher reflection to discuss your current use and facilitation of small-group discourse during your lessons.

> **TEACHER** *Reflection*
>
> During small-group discourse, what types of instructional questions do you use to support student discussion and perseverance? What are strategies you use to support perseverance through the completion of a mathematical task?
>
> _____
> _____
> _____
> _____
> _____
> _____
> _____
> _____
> _____
> _____
> _____
> _____
> _____
> _____
> _____
> _____
> _____
> _____

How can you support student discourse during small-group instruction? First, you leave the front of the room and walk among the student teams. You listen in on their discussions, mistakes, and possible solution pathways. Second, you understand that your role is to provide feedback. You are to facilitate the small-group classroom conversations and support student listening, engagement, and learning. You also learn quickly to provide feedback prompts that ensure all students have a chance to learn—not just the students who are first to talk in their group. Figure 10.8 (page 88) has K–12 sample prompts you can use to support students while maintaining engagement. At the bottom of the chart there is room for you to also note other prompts you might use.

Use the questions to reflect on your use of assessing and advancing questions during small-group discourse.

When introducing and using a task during small-group work, it's important to set up clear expectations of how you want students to engage with each other and in the task. Then, as you walk around and support students, you can utilize the assessing and advancing prompts to help each student access the task.

When students make errors during the lesson, they need to receive FAST feedback on those errors, and then take action to correct their mistakes in reasoning. Students then view reflection and refinement of their in-class work as something they *do* in order to focus their energy and effort for future learning. As indicated so far in this section, a great place for this type of student reflection is during small-group discourse, as your students work together on various problems or tasks you provide, and you walk around the room, evaluate understanding based on what you see *and* hear, and then provide meaningful feedback to students and student teams. This process turns your checks for understanding into more meaningful formative assessment processes.

Thus, you and your students share the responsibility for successful implementation of in-class formative assessment practices. When your students can demonstrate *understanding through reasoning mathematically*, they connect to the essential learning standard for the unit and can reflect on their individual progress toward the learning target of that day's lesson. You support students' progress by using immediate and effective feedback during the daily classroom conversations, whether in whole-group or small-group moments, and then expecting your students to act on the feedback provided.

To support the FAST formative feedback process described on page 62 using small-group task exploration, you will need scaffolding (or "unstucking") prompts and advancing prompts ready to go for each task of the lesson. *Unstucking prompts*, also known as *assessing prompts*, are questions or statements you use to help students access the content or start the task if they are stuck. *Advancing prompts* are questions or statements you use to extend a task for a group of students who are demonstrating understanding on the current task. You will need to think of both the assessing and advancing prompts you will use with students before you implement the actual mathematics task with students in class.

Assessing and Advancing Prompts to Ask Students During Small-Group Discourse	
Prompts that help students work together to make sense of mathematics: • Who agrees? Disagrees? Who will explain why or why not? • Who has the same answer but a different way to explain it? • Who has a different answer? What is your answer, and how did you get it? • Can you please ask the rest of the class that question? • Can you explain to your partner your understanding of what was just said? • Can you convince us that your answer makes sense?	Prompts that help students learn to reason mathematically: • Does that always work? Why or why not? • Is that true for all cases? Explain. • What is a counterexample for this solution? • How could you prove that? • What assumptions are you making?
Prompts that help students learn to conjecture, invent, and solve problems: • What would happen if _____? What if it did not happen? • Do you see a pattern? Explain. • What about the last one? • How did you think about the problem? • What decision do you think he or she should make? • What is alike and what is different about your method of solution and his or hers? Why?	Prompts that help students connect mathematics, its ideas, and its applications: • How does this relate to _____? • What ideas that we have learned before (prior knowledge) were useful in solving this problem? • What problem have we solved that is similar to this one? How are they the same? How are they different? • What uses of mathematics did you find in the newspaper last night? • What example can you give me for _____?
List other questioning prompts you use:	

Source: Adapted from NCTM, 2009.

Figure 10.8: Team discussion tool—Assessing and advancing questions to ask students.

Visit **go.SolutionTree.com/MathematicsatWork** *for a free reproducible version of this figure.*

Monitor Actions and Results

You can use figure 10.9 as a tool to collect information and provide feedback on the actions and results of your students and their team engagement in the mathematical tasks and activities you plan for class. Ask a few fellow teachers to come in and help you with these observations as you need them.

As part of this formative process, you provide guidance and scaffold questions to support student learning and perseverance on the task. You also determine if you

---

TEACHER *Reflection*

Refer to figure 10.8. How could you use these types of questions within your own classroom?

What are one or two questions you could try using tomorrow?

How will you ensure all students take action on your feedback during their work together on a mathematics task?

	1—Working task but constantly stuck as a student team; does not generally take correct action on the feedback and prompts from the teacher	2—Working task through connecting to prior knowledge, engaging in conversations, and taking action to the scaffolding prompts from the teacher	3—Working task through engaging in accurate sense-making and reasoning, using multiple connections, and minimal feedback from the teacher	4—Working task is correctly done, and student team engages in an extension (advancing) prompt from the teacher during the small-group discourse	Next steps for the student team based on teacher observations
Student team 1					
Student team 2					
Student team 3					
Student team 4					
Student team 5					
Student team 6					
Overall observations of student reasoning and engagement:					

Figure 10.9: Team discussion tool—Walk-around formative assessment.

*Visit **go.SolutionTree.com/MathematicsatWork** for a free reproducible version of this figure.*

need to make adjustments in your whole-group instruction to develop students' reasoning skills. If a student team is ready, you encourage it to try the extension prompt or activity for the task as well.

You and your collaborative team should discuss how this tool or a modification of it might be useful in providing information about students' development of proficiency to reason abstractly and quantitatively during your mathematics lessons.

As students are wrapping up their work on a task and it is time for you to select students or teams who will present the solution process to the class, follow these three steps Smith and Stein (2011) outline.

1. *Select* particular students to present their mathematical work during the whole-class discussion.

2. *Sequence* the student responses that you will display in a specific order.

3. *Connect* different students' responses and connect the responses to key mathematical ideas.

As you are selecting students to present (either randomly or based on their specific pathway), remind students to use *we* when explaining their team's solution (not *I*) as it helps to build the classroom as a community. Also think about how you can tell a story with the different strategies and solutions students produced during small-group discourse time. This is when the idea of sequencing student responses might come into play. The sequence can tie the mathematics together and allow you to connect the learning back to the essential learning standard for the lesson.

Note that in this "wrapping up the task" process, you do not go over the problem, rather you allow student work and answers to be highlighted. You might assign a specific student as an expert for all similar problems of the same type or standard in the future.

Task implementation within small-group discourse, to be successful, will require you to manage certain classroom structures every day. This is the focus of the final section of this chapter.

Manage Student Teams

Managing students in peer-to-peer productive discourse is critical for effective implementation of lower- and higher-level-cognitive-demand tasks. When you set your student teams to work, do they know the expectation for teamwork or do they rely on one or two students to get them started? Do they understand their rights and responsibilities? Do they know norms for behavior and how to work effectively in small groups? Your students will need structure to support meaningful mathematics discourse, engagement, and action.

Your collaborative team can create posters to share with students, or you can create these with your students. Figures 10.10 and 10.11 illustrate examples. Specifically, figure 10.10 offers rights and responsibilities for classroom discourse. Figure 10.11 offers norms of behavior and skills for small-group learning.

Before you can achieve access to the instructional learning targets for your lesson, it is necessary to ensure a safe classroom climate with clear expectations for student sharing and behavior. Students need to feel comfortable sharing their ideas and taking risks in front of their peers. There is a benefit to EL and special education learners discussing ideas with their team members first before sharing their thoughts more publicly with the entire class. Take a few moments to reflect on your current norms and expectations for your students.

> **TEACHER *Reflection***
>
> How do you currently create and then use class norms and expectations to help your students facilitate the sharing of ideas?
>
> _____
> _____
> _____
> _____
> _____

Rights	Responsibilities
Each student has the right to: • Ask for help • Be heard • Make a mistake • Express his or her thoughts on solution pathways • Learn the standard • Disagree with respect	Every student should: • Help others when asked by teacher or peers • Listen to the ideas of others • Take action on feedback • Be open to embracing errors • Seek consensus within his or her team • Work toward success with his or her peers

Figure 10.10: Sample classroom discourse rights and responsibilities.

Visit **go.SolutionTree.com/MathematicsatWork** *for a free reproducible version of this figure.*

Norms of Behavior	Skills for Small-Group Learning
Student team members should: • Listen carefully and with respect to one another • Contribute to the assigned team task • Ask other team members for help when needed • Help other team members who ask • Insist on logical persuasion before changing your mind	Student team members should: • Use quiet-conversation-level voices • Stay engaged and persevere through the mathematical task assigned • Ask peers for help, and then ask the teacher • Be supportive of each other • Ask for reasons or ask each other questions • Criticize ideas, but not the other students on your team • Have a sense of humor

Figure 10.11: Sample classroom norms.

Visit **go.SolutionTree.com/MathematicsatWork** *for a free reproducible version of this figure.*

One way to engage students in the process of developing norms for collaboration is to invite your students to respond to the following.

- Ask your students how they try to disagree with someone in a nice way.
- Discuss with your students what it means to make the conversations about mathematical ideas—and not about the person.
- Ask your students how team members should respond when someone on the team isn't participating.
- Ask student teams for strategies they can use as a team before they need to involve your support.

There are several cooperative learning structures you and your colleagues can employ to create required participation, the key to engaging students in mathematical thinking—whatever the structure or activity (Johnson & Johnson, 1999; Johnson, Johnson, & Holubec, 2008; Kagan, 1994; Kagan & Kagan, 2009). Each of the following structures requires the use of a seating chart. See figure 10.12 for a sample seating chart template. Consider the following four strategies.

1. **Use a structured seating chart:** This strategy makes it efficient to randomly call on a group to share or to call on specific students within a group. Since each student team has an assigned number, and each of the four students on the team is numbered from one to four, you can roll a die and call on student three from team five to present his or her team's solution. You can also use the structure to quickly organize the student work. For example, you can launch the mathematical task and state, "Student four in each team will lead the discussion when I give the signal."

2. **Use group and seat numbers to assign roles:** For example, all students who are number threes read the problem while number twos and fours lead the discussion, and the number ones write down the solution being discussed.

3. **Randomly choose which student's paper you will collect:** This structure is one way to ensure all students keep on pace together and don't work ahead of other team members. A key factor to student team success is making sure everyone on the team corrects his or her errors and understands a solution pathway to the mathematics task. If the students do not know which paper will be collected, they are

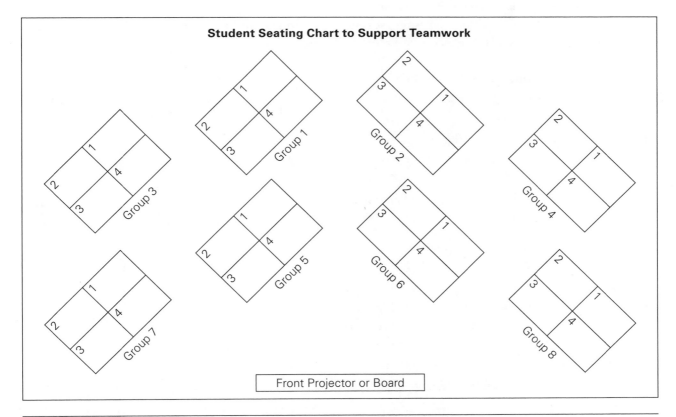

Figure 10.12: Seating chart to support the work of student teams during instruction.

Personal Story — TIMOTHY KANOLD

I was working in a suburban Minneapolis school district for over seven months helping its teachers move out of rows and into teams of four for student discourse. At one point, I walked into a mathematics classroom of a teacher who had had a lot of success using the new classroom arrangement of desks to facilitate small-group discourse. Imagine my surprise to see her students sitting in seven rows of five deep!

I sat in the back, initially dismayed, but said nothing as I prepared to take notes on the student-engagement levels during the lesson. The bell rang, and all of the students had a worksheet on their desks. On the teacher's command, they got started on the first math problem on the worksheet. I asked her for a copy and noted that it was a vertically stacked set of five questions, meaning that success in the five questions was somewhat dependent on understanding and success in each previous question?

But then the class got exciting! On her command, the students stood up, left their papers on the desk, and rotated back one desk and chair. The person in the back rotated to the front of his or her row. The students were allowed to fix the solution shown currently in question one, and then they moved on to question two. These rotations continued until each student returned to his or her original paper where all five solutions awaited. They then had one last chance to reflect on the feedback they received and edit any parts of the five solutions.

Students turned in their papers and moved their desks back into teams. Then they turned to me and with their teacher thanked me for helping her make class a more exciting place to learn math. Go figure!

---

more likely to ensure understanding for each student on the team.

4. **Write notes on the seating chart:** This tool can also help when recording scores or tracking student participation and collecting data for work completed and perseverance levels. It can also help to remind you of a student observation you made during the lesson.

The seating chart is a helpful tool for capturing data as well. For example, while observing your class, a colleague could use the following coding system to capture the type of questioning interactions within your class.

- Circle (o) = Student raises hand to ask question.
- Asterisk (*) = Student calls out answer.
- Square (□) = Teacher calls on student who doesn't have hand raised.

You could also use the seating chart to help track your movement around the room during small-group discourse. Use different colors depending on the type of small-group interaction you observe. For example, when you stop at a student group, use the following indication.

- Red = Supports the team by asking unstucking or scaffolding questions
- Green = Quietly observes student discussion and continues to the next group
- Blue = Supports the group by asking advancing questions for the mathematical task

Take a few moments to reflect on how you manage your student teams during your lessons.

TEACHER Reflection

How do you currently manage your student teams in your classroom? Do you use teams of four? What strategies and structures do you intentionally plan to help create a safe culture for sharing and risk taking by your students and their peers?

_____
_____
_____
_____

As a team, use the tool in figure 10.13 to reflect on your current effort to manage your student teams.

Directions: As a team, discuss and share your ideas and strategies about management of student teams during the lesson.

1. How do you currently organize your teams of four?

2. Think about your small-group or team activities, such as warm-ups, tasks, or closure activities. What type of management commands and prompts do you give student teams before they go to work?

3. What do you do while your teams are working? What behaviors do you reinforce? How do you provide feedback if students are stuck? How do your teams know how to behave (norms)?

4. Once the teams finish and you are ready to return to a whole-group focus, what verbal or nonverbal commands do you use?

5. Once you have the whole-group focus, how do you summarize the problem or task with the whole group? Do you do it? Do you ask students to do it?

6. List any techniques you use to coach your students to listen to each other during whole-group discourse.

7. Are there any other team activities you feel you do that support successful small-group discourse and learning?

Figure 10.13: Team discussion tool—Management of student teams.

*Visit **go.SolutionTree.com/MathematicsatWork** for a free reproducible version of this figure.*

As you and your team discuss your responses to the questions in figure 10.13 (page 93), you may be unsure how to respond, especially if implementing teams of four is new to you. That is okay.

You can then use figure 10.14 to continue your teacher team discussion and reflect on next steps with your student teams. It contains responses to the questions in figure 10.13.

Directions: Possible responses to the questions from the management of student teams discussion tool (figure 10.13, page 93) appear below. Highlight one or two responses in each category that you would like to start trying in your own classroom.

1. How do you currently organize your teams of four?

   It is imperative to not mix the students homogeneously during instruction because this can cause greater gaps in learning. Use only a heterogeneous mix for each team of students.

2. Think about your small-group activities, such as warm-ups, tasks, or closure activities. What type of management commands and prompts do you give student teams before they go to work?
   - Nonverbal commands such as turning the lights out or raising your hand and expecting all students to raise their hand when they see you do it
   - Time commands such as, "You have five minutes, get started now."
   - Specific commands such as, "Work on the warm-up silently for one minute, and then share with your team."
   - Sharing commands such as, "Twos, you're the captains. After three minutes, ask each member of your group how to do the problem."
   - Timed directions such as, "In your groups, discuss for three minutes."
   - Desks out for silent work, and desks turned back in for working together
   - Directions such as, "Get started on the warm-up," with the invisible norm that students are to work alone and then pair-share
   - Visual directions—not just auditory—to support the different learners in your classroom

3. What do you do while your student teams are working? What behaviors do you reinforce? How do you provide feedback if students are stuck? How do your teams know how to behave (norms)?
   - Extensively tour the room and observe student work.
   - Redirect questions back to the whole group.
   - Tour student groups and listen to student conversations.
   - Tour and reinforce success and encourage those who are not working with comments such as, "Right idea!"
   - Listen and kneel to notice work that is done well, and tell students to share with someone who is struggling with the problem.
   - Positively reinforce students who are helping other students and encourage them.
   - Keep track of time and remind students with a verbal command such as, "You have two minutes left."

4. Once the teams finish and you are ready to return to a whole-group focus, what verbal or nonverbal commands do you use?
   - Nonverbal commands such as walking back to the front of the room, placing a hand in the air for students to focus on, or turning lights back on
   - Verbal commands that might include:
     a. "Pencils down."
     b. "Okay, eyes back up front, pencils down, and listen to me."
     c. "Can I have your attention up front?"
     d. "Boys and girls" or "Ladies and gentleman"
     e. "I like what I am seeing from group _____ and group _____. They have their eyes up front, and they have stopped their conversation."
   - Whether using verbal or nonverbal commands, wait until you have everyone's attention to talk to the whole group.

5. Once you have the whole-group focus, how do you summarize the problem or task with the whole group? Do you do it? Do you ask students to do it?
   - Tour the classroom to gauge everyone's understanding of the task you assigned so there is no need to go over the problem.
   - Have the students summarize; for example, all the threes go to the board to share (or share from their devices), a pair of students presents the solution at the board, or a randomly selected student presents the problem.

Figure 10.14: Team discussion tool—Management of student teams (with answers).

	• Ask one student who has the correct answer or multiple students with different strategies and the correct answer to put his or her solution on the board as you continue touring the classroom and helping small groups with the problem. • Uncover the solution on the board and ask students to compare their small-group solutions with the one you present. What's similar about their strategy, and what's different? • Have groups write out their group solution on a piece of paper to present using a device in front of the class. • Have students show their work on whiteboards or chalkboards around the room so everyone can see all student work and the group can quickly examine and discuss any solution as it needs to.
6.	List any techniques you use to coach your students to listen to each other during whole-group discourse. • Explain to students that you expect them to listen to each other's responses. • Redirect questions back to other students in the class, and use a teacher explanation only when no other student can explain the solution or the reason for the solution. • Often restate the question or refuse to answer any student's question immediately, and try to allow other students to think about the question. • Rephrase the question if someone asks a question and then ask the question of the whole class. • Make the question sound important by saying, "That's a great question, who could answer this question or help _____ out?" Redirect the question to the class. • If an individual student asks a question, and it's important for the whole class to hear the question, state the question back to the whole class to think about a response. Possibly even have students discuss the question in their groups before answering in the whole group. • Eliminate student callouts by qualifying student questions with phrases such as, "Take thirty seconds to think quietly" or, "Raise your hand." • Record what teams say in terms of the mathematics and the social skills they demonstrate as they work on a task. Share these with the class as a celebration for great mathematics talk.
7.	Are there any other team activities you feel you do that support successful small-group discourse and learning? • Use a student-evaluation form to gather information from teams about individual student participation to help support teams. • Plan activities in advance of the lesson to ensure they go well. • Do timed team challenges for friendly competition in class, which adds energy to class—especially on days when students may seem sluggish.

Visit go.SolutionTree.com/MathematicsatWork for a free reproducible version of this figure.

As you reflect on the strategies in figure 10.14, you will likely realize it may be difficult to implement all the strategies immediately. However, it is important to start to move toward implementation of *teams of four* while also implementing strategies and structures to support a safe environment for students to take risks and share ideas during small-group discourse.

Promoting productive small-group student discourse will allow your students to find their voice in your lesson discourse and actively engage. Students can self-assess how they are working together as you provide feedback to student teams on their ability to work cooperatively (Johnson & Johnson, 1999; Kagan, 1994; Kagan & Kagan, 2009). Figure 10.15 (pages 96–97) presents just such a student self-evaluation form you might use in class.

Figure 10.15 provides information on the collaborative process from the student perspective. You and your team can easily adapt this tool to an online form or survey where you ask students to reflect and respond.

This chapter has focused on your mathematics task and discourse implementation. The student management necessary to effectively transition from one type of discourse to the next will allow you and your students to more effectively respond to PLC critical questions three and four: What will be your response in class when students are not learning, and what will be your response in class when students are learning?

This is the exact reason to balance the use of small- and whole-group discourse. To respond to PLC critical questions three and four, you must first know if they are learning during the lesson. That is impossible *if* you

Evaluator name: _____		
Class: _____		Group number: _____

Person 1	Name:	
This person is willing to help and explain when asked.		Never—0 Sometimes—1 Always—2
This person does his or her share of group work.		Never—0 Sometimes—1 Always—2
This person gives logical explanations as opposed to saying, "Just because" or "Believe me."		Never—0 Sometimes—1 Always—2
This person contributes to learning and makes it fun to work together.		Never—0 Sometimes—1 Always—2
Write the sum of the four scores for person 1 in the blank. (If person 1 is you, please circle the sum.) Sum = _____		
Person 2	**Name:**	
This person is willing to help and explain when asked.		Never—0 Sometimes—1 Always—2
This person does his or her share of group work.		Never—0 Sometimes—1 Always—2
This person gives logical explanations as opposed to saying, "Just because" or "Believe me."		Never—0 Sometimes—1 Always—2
This person contributes to learning and makes it fun to work together.		Never—0 Sometimes—1 Always—2
Write the sum of the four scores for person 2 in the blank. (If person 2 is you, please circle the sum.) Sum = _____		
Person 3	**Name:**	
This person is willing to help and explain when asked.		Never—0 Sometimes—1 Always—2
This person does his or her share of group work.		Never—0 Sometimes—1 Always—2
This person gives logical explanations as opposed to saying, "Just because" or "Believe me."		Never—0 Sometimes—1 Always—2

This person contributes to learning and makes it fun to work together.	☹ 😐 🙂 Never—0 Sometimes—1 Always—2
Write the sum of the four scores for person 3 in the blank. (If person 3 is you, please circle the sum.) Sum = _____	
Person 4	**Name:**
This person is willing to help and explain when asked.	☹ 😐 🙂 Never—0 Sometimes—1 Always—2
This person does his or her share of group work.	☹ 😐 🙂 Never—0 Sometimes—1 Always—2
This person gives logical explanations as opposed to saying, "Just because" or "Believe me."	☹ 😐 🙂 Never—0 Sometimes—1 Always—2
This person contributes to learning and makes it fun to work together.	☹ 😐 🙂 Never—0 Sometimes—1 Always—2
Write the sum of the four scores for person 4 in the blank. (If person 4 is you, please circle the sum.) Sum = _____	

Figure 10.15: Student team processing-and-evaluation tool.

Visit *go.SolutionTree.com/MathematicsatWork* for a free reproducible version of this figure.

do not get away from the front of the room and you cannot hear what the students are thinking.

In this final story, author Jessica Kanold-McIntyre relates an early experience as a middle school mathematics teacher.

Thus, the best closure prompts and activities come in the form of information *from students* about what they learned during the class—for example, a restatement of the instructional purpose. This information then provides knowledge of the results for the teacher; did

Personal Story 〰 JESSICA KANOLD-McINTYRE

I was doing my student teaching in the spring of 2004. I had some battles with my cooperating teacher about using teams of four to create an improved student-engaged climate for learning. She wanted the students to stay in rows as she did all of the work from the front of the room. We did battle.

It would have been great to have the 2009 research by John Hattie at the time, where he provides a great reminder of what matters most in teaching: *The role of the teacher is to constantly assess and have a pulse on where each and every student is in relation to the essential learning standards for the unit.*

So many times in my classroom I would jump in to summarize or "help" a student, which meant I would do all the talking instead of all the listening. I had to learn to let my students do the talking. I love that Hattie, now in 2018, validates my early understanding of good mathematics instruction.

TEAM RECOMMENDATION

Task and Discourse Implementation

Consider the following recommendations with your team.

- Know that the cognitive demand and rigor of the tasks you choose matter in order to promote high-quality, small-group discourse.
- Comprehend that small-group discourse allows the formative feedback process to come alive in your classroom.
- Consistently have your students in teams of four to establish a culture of collaboration and sharing. Take time to establish routines and expectations for how to hold a discussion and listen to each other.
- As students work and share ideas in groups, be sure to circulate and listen to student conversations so you can ask advancing or assessing questions and know how to sequence the whole-group summary of the task.
- Understand how the classroom culture and climate impact students' sense of safety for sharing, which in turn impacts student learning and willingness to take risks.
- Establish classroom norms for whole-group and small-group behavior.
- Consider having students give you feedback on how comfortable they feel sharing and collaborating in your classroom. You could create a simple online survey for students to express their thoughts.

you teach what you intended to teach, and have the students learned what you intended to have them learn?

As you reflect on your instructional practices, eventually the time element to the lesson comes into play, especially as you use more small-group discourse as Jessica did. You will not be able to cover quite as much mathematics content "ground" in each lesson. Yet, class must end. There is one last element to consider within a classroom environment that fosters student perseverance and engagement. This is the way you bring the lesson home—lesson closure, as the final evidence of student learning that day.

CHAPTER 11

Using Lesson Closure for Evidence of Learning

Students learn when they are encouraged to become authors of their own ideas and when they are held accountable for reasoning about and understanding key ideas.

—Randi A. Engle and Faith R. Conant

As you and your team wrap up your lesson, how do you know if students understand the learning target for the day? What evidence will you collect, and what do you still need to ask in order to know where each and every student is in his or her understanding of the lesson learning targets?

In *Visible Learning*, Hattie (2009) writes:

> It was only when I discovered that feedback was most powerful when it is from the *student to the teacher* that I started to understand it better. When teachers seek, or at least are open to, feedback from students as to what students know, what they understand, where they make errors, when they have misconceptions, when they are not engaged—then teaching and learning can be synchronized and powerful. Feedback to teachers helps make learning visible. (p. 173)

Being *intentional* in your planning to include a closure activity at the end of every lesson—the final lesson-design element—is another way to embed formative assessment into your daily lesson plans.

Facilitating Closure Activities

Closure activities should be student led so you can ensure students understand the learning targets of the lesson. If students' current conceptions are incorrect or they have misunderstandings, a student-led summary as a closure activity gives you great feedback on next steps in future lesson decisions.

How do you currently facilitate a student-led summary? Should it be whole-group, small-group, or individual activity? What does it look like in the classroom? What would you see or hear students doing during this time? Author Mona Toncheff provides some insight into the importance of this ending aspect of your lesson in her story on page 100.

Figure 11.1 (pages 100–101) presents possible closure activities that allow you to address student learning from the lesson while also keeping the focus on a *student-led* summary of the lesson rather than a teacher summary. These types of activities provide great data about where students are in their progression of understanding the essential learning standard. They provide solid evidence for you and your team in your planning process.

As you and your team members think about how you want to implement lesson closure for evidence of student learning, remember there are many ways you can have students lead the closure activities. The key is that students are *doing* the reflection and summary around the learning target for the day, and that it is not teacher led. Lesson closure provides an informal way to take the pulse on student understanding. Student-led closure provides you with critical information about how effective your lesson was at supporting student learning of the daily learning target that aligns to the essential learning standard of the unit. In his story on page 101, author Tim Kanold describes his early experiences on the issue of closure with his mathematics articulation committee.

> ## Personal Story 〞 MONA TONCHEFF
>
> I will never forget my first coaching experience when I was a new teacher. The instructional specialist came into my classroom to observe me as a first-year teacher. She provided positive feedback on student engagement and then asked me one simple question, "How do the students know that they met the learning target for the day?" At that point in time, I didn't have a response.
>
> As I reflected on the lesson, I was thinking about the evidence of student engagement as students were working effectively with their peers to make sense of the mathematics. The students looked like and sounded like they met the target, however, I had not given the students an opportunity to reflect on their own learning. I assumed students would individually take responsibility to reflect on the learning for the day. What I didn't realize as a new teacher is that not all students enter into my classroom as reflective learners.
>
> During my first five years in the classroom, I had to work diligently to give students time at the end of each lesson to wrap up their learning for the day so that when they left my classroom, they could articulate what they spend their class time learning. When I became more fluent in closure strategies, my students also grew into reflective learners, asked better questions of their peers, and understood how to use the learning targets as a pathway for their learning.

Activity	Description	How Activity Is Formative
Student-reflection exit slip	Use a specific question that ties to the content from the class. The question is of higher cognitive demand to assess true understanding at a conceptual level rather than just a procedural level.	For the exit slip to be formative, there must be teacher and student action on the information. For example, the teacher must review answers, sort the results into groups (got it, almost got it, not yet), and then give each group a specific problem to begin the lesson the next day.
Student team summary	Have groups write or draw what they learned for the day and share with the class.	For this activity, students who are listening to the summary should ask the group questions and provide feedback—like a Socratic discussion. The feedback to students is immediate, and the teacher can document group understanding to use in her planning for the next day.
Questioning—small group or whole group	Ask students questions such as: • "Why?" • "Could you explain it another way?" • "How does this connect with _____?"	The questions must be crafted to facilitate a conversation that provides feedback on student understanding. Through the questioning process, the feedback is immediate to students, which helps them shape their understanding in the moment. As a teacher, you are responding formatively by listening and choosing your next question based on the answers from the students.
Gallery walk	Capture the complex task students worked on in class on poster paper and hang the paper around the room. Students walk around the room giving feedback such as: • "I wonder . . ." • "I like . . ."	After the gallery walk, the class provides feedback and students make adjustments to their work based on the feedback (that day in class or the next day).

Student presentations	Have student groups present their work from a task from class or present their summary of the lesson.	During the presentation, students record specific content each group mentions and offer a note about one thing they like and one question they have. The groups get this feedback to review and adjust their thinking.
Voting with feet	Pose agree or disagree questions and ask students to move to one side of the room or the other depending on whether they agree or disagree.	What would normally be a check for understanding can turn into a fun classroom debate between differing sides by having students explain why they chose their answer. Students then revote after the debate.
Nonverbal check	Using a scale of 1–5, ask students to hold up the appropriate number of fingers to their chest to indicate their comfort level and confidence with the learning target for the lesson. (Alternatively, you can use thumbs up or thumbs down.)	Prepare multiple questions and have them ready to go as a check-in with students. After the nonverbal check, regroup students for a re-engagement activity based on self-reported responses. In those new groups, provide students differentiated instructional tasks with specific feedback as needed.
Voting tools (like Google Forms, Schoology, Edmodo, Haiku Learning, Go Formative, and so on)	Give online quizzes where students get their results immediately and you can see all student results.	This is a great way to capture actual data for each and every student in an efficient way, but it can be difficult to make feedback formative. Some tools allow the teacher to type a response directly back to the student. The data can also be used to regroup students for a differentiated warm-up activity the next day.
Online discussion forums (like Schoology, Google Classroom, Edmodo, Socrative, TodaysMeet, The Backchannel, and so on)	Have students participate in online classroom discussions where they share their thinking, read classmate explanations, and learn from each other.	This strategy is a great way to use technology to provide students with a forum to communicate about their mathematics learning outside of the classroom. Provide specific questions tied to essential learning standards at the end of a class, or use the forum as a way for students to ask each other questions about homework, and so on. This requires clear expectations for student behavior and some monitoring by the teacher, but it can provide positive support.

Figure 11.1: Team discussion tool—Sample lesson-closure activities.

Visit **go.SolutionTree.com/MathematicsatWork** *for a free reproducible version of this figure.*

Personal Story — TIMOTHY KANOLD

At Adlai Stevenson, four to five times a year we would gather our Mathematics Articulation (MAC) Task Force. As the director of mathematics, it was my responsibility to chair these professional development events as we tried to bring a K–12 coherence to our mathematics efforts. After several years of meeting together (schools often sent different teacher representatives each year), we realized our K–12 mathematics lessons were "incomplete stories."

We had become pretty good at starting the story of the lesson for each day. By then we had made a K–12 commitment to students learning mathematics in teams of four, even buying new horseshoe-style desk furniture with separate student chairs. We were also good at opening the story with warm-up activities and connections to the learning standard (back then we called them SLOs—student learning objectives) for the day.

continued →

To some extent the middle of our mathematics lesson story was pretty good, too. We were becoming much better as teachers of mathematics at using meaningful mathematics tasks and engaging the students in discussions. We did not have a good ending to the story. Ever. As a body of K–12 teachers, we just thought if the students seemed to be doing okay, then they knew what they were doing. There was no need to end the lesson's story.

To prepare for my meeting with the MAC, I spent two months traveling from school to school and observing teachers of mathematics. I would ask students one question at the end of the lesson (in elementary) or end of the class period (in middle school and high school): "In thirty seconds or less, tell me what you learned today?" Not surprisingly, even in the lessons that went extremely well, the students' responses were widely different, sometimes inaccurate, or inarticulate. They just did not know the story of the mathematics lesson.

As teachers, it was our fault. We were not providing any opportunity for the story to end and for the students to own the ending as they reflected on their work for the day. I gathered several teachers from different grade levels, and the five of us (I was teaching a class of calculus AB) started experimenting in our classes on various endings. We intuitively knew that how students feel (level of confidence) was related to how they could summarize what they knew from that day's lesson. We asked students to keep journals and write their responses to our prompts, and then share with a partner. These were simple student reflection questions, not a complicated set of exercises.

When our MAC Task Force met, we submitted our findings, each of us relating our experiences and our surprise that our students did not remember or view understanding of the mathematics lesson the same as we did. We began our journey of making sure the story of the mathematics lesson would always have an ending they would write.

Reflecting on Effectiveness

Not only is it important to plan for and implement student-led closure, it is also important to personally reflect on the effectiveness of the lesson with colleagues if possible. The lesson-closure activity can be very helpful to your lesson-reflection process. Notice that in the Mathematics in a PLC at Work lesson-design tool discussed in part 1, chapter 7, figure 7.1 (pages 52–53), and shown again now in part 2 as figure 11.2, there is a space at the bottom for your personal end-of-lesson reflection. Leave notes for yourself for the following lesson, unit, or year and reflect on next steps in your instructional unit plan. Specifically, what might need to change in tomorrow's lesson based on student evidence of learning from *today*?

Author Sarah Schuhl shares her experience helping a teacher team use the Mathematics in a PLC at Work lesson-design tool on page 105. Re-examine each of the criteria as you consider the components of your current mathematics lessons.

Now that you and your team have reflected on the important features of a lesson and how to implement each lesson-design strategy to support student perseverance and formative assessment, visit **go.SolutionTree .com/MathematicsatWork** to consider an appropriate sample lesson. You will find a sample grade 1 lesson plan, a sample grade 5 lesson plan, a sample grade 8 lesson plan, and a sample high school advanced algebra lesson plan.

As you examine the sample lessons online, remember that these sample templates are to help support your planning effort and focus for the mathematics content and processes of any daily lesson. Do not approach the tool as something that must be completely filled in as much as a guide for the types of student-engaged activities you and your team need to plan for as you

Preparing for the Lesson
Unit: Fill in the title of the unit. **Date:** Fill in the date of the lesson. **Lesson:** Provide a short descriptor about the nature of this lesson.
Essential learning standard: State the essential content and process standard *for the unit* you address during *this* lesson. • **Content**—Write as an *I can* statement. • **Process**—Write as an *I can* statement.
Learning target: State the specific learning outcome(s) for this lesson. *Use, "Students will be able to . . ."*
Academic language vocabulary: State the academic vocabulary expectations for the lesson. Describe how you will explicitly address any new vocabulary.

Beginning-of-Class Routines
Prior knowledge: Describe the warm-up activity you will use. How does the warm-up activity connect to students' prior knowledge, connect to an analysis of homework progress, or connect to future learning?

During-Class Routines
Task 1: Cognitive Demand (Circle one) *High* or *Low* What are the learning activities to engage students in learning the target? Be sure to list materials as necessary.

What will the teacher be doing?	What will the students be doing?
• How will you present and then monitor student response to the task? • How will you expect students to demonstrate proficiency of the learning target during in-class checks for understanding? • How will you scaffold instruction for students who are stuck during the lesson or the lesson tasks (assessing questions)? • How will you further learning for students who are ready to advance beyond the standard during class (advancing questions)?	• How will you actively engage students in each part of the lesson? • What type of student discourse does this task require—whole group or small group? • What mathematical thinking (reasoning, problem solving, or justification) are students developing during this task?

Task 2: Cognitive Demand (Circle one): *High* or *Low*
What are the learning activities to engage students in learning the target? Be sure to list materials as necessary.

Figure 11.2: Mathematics in a PLC at Work lesson-design tool.

continued →

What will the teacher be doing?	What will the students be doing?
• How will you present and then monitor student response to the task? • How will you expect students to demonstrate proficiency of the learning target during in-class checks for understanding? • How will you scaffold instruction for students who are stuck during the lesson or the lesson tasks (assessing questions)? • How will you further learning for students who are ready to advance beyond the standard during class (advancing questions)?	• How will you actively engage students in each part of the lesson? • What type of student discourse does this task require—whole group or small group? • What mathematical thinking (reasoning, problem solving, or justification) are students developing during this task?

Task 3: Cognitive Demand (Circle one): *High* or *Low*
What are the learning activities to engage students in learning the target? Be sure to list materials as necessary.

What will the teacher be doing?	What will the students be doing?
• How will you present and then monitor student response to the task? • How will you expect students to demonstrate proficiency of the learning target during in-class checks for understanding? • How will you scaffold instruction for students who are stuck during the lesson or the lesson tasks (assessing questions)? • How will you further learning for students who are ready to advance beyond the standard during class (advancing questions)?	• How will students be actively engaged in each part of the lesson? • What type of student discourse does this task require—whole group or small group? • What mathematical thinking (reasoning, problem solving, or justification) are students developing during this task?

End-of-Class Routines

Common homework: Describe the independent practice teachers will assign when the lesson is complete.

Lesson closure for evidence of learning: How will lesson closure include a student-led summary? By the end of the lesson, how will you measure student proficiency and that students develop a deepened (and conceptual) understanding of the learning target or targets for the lesson?

Teacher end-of-lesson reflection: (To be completed by the teacher after the lesson is over)
Which aspects of the lesson (tasks or teacher or student actions) led to student understanding of the learning target? What were common misconceptions or challenges with understanding, if any? How should you address these in the next lessons?

Visit *go.SolutionTree.com/MathematicsatWork for a free reproducible version of this figure.*

Personal Story — SARAH SCHUHL

To help a second-grade team more fully understand shared teacher learning, we planned a lesson together for two-step word problems, an important concept for students to learn and one that is often difficult to teach. The teachers identified the standard, essential learning standard, and daily target to frame the lesson. In the discussion, one teacher realized that the emphasis is not on two steps, but rather that students reason and think through the task to its completion. Some students might think about the problem in two steps, and others might complete a step mentally and show one complete step. The planning idea was that students would start and complete the task in different ways and then be able to share their solution strategies and answers.

Armed with this information, we talked about what students have learned prior to the lesson that will give them the tools and resources to work through the two-step tasks. Students had solved one-step word problems with the unknown in all parts of an equation and added and subtracted within 100.

The team decided to launch the lesson with a prior-knowledge activity using a missing addend word problem that required double-digit addition or subtraction to solve. Students would work together in groups to share ideas for solving the task. Next, we identified the academic language students and teachers needed to intentionally use in the lesson with sentence frames or on a word wall for clear discourse in class.

The teachers then identified three tasks from their curriculum to use for the lesson—one that would be modeled using their problem-solving template (What do you know? What do you need to know? Draw a picture to explain the problem. Show your thinking to solve the problem. Write your answer.) They would then ask students to work in pairs to solve the next task and set a timer to keep students focused.

After two minutes, they would have students mix up and share their starts with one another for feedback before going back to their original partner and working or modifying their solution. After another five minutes, pairs would share their thinking in their original teams and then share selected solutions with the class.

The third task would be completed by groups of four and posted around the classroom. Teachers carefully thought about how students would use discourse and visible thinking to give feedback to one another while the teacher also could be available to work with students and groups for formative feedback.

One teacher realized that she had only been modeling word problems and then having students work on them, not teaching students to see each other as resources and learning partners. Throughout the discussion, we also talked about how to differentiate the learning by being prepared with a manipulative for students who might struggle and extending the learning by asking students to *create* a word problem requiring a student to add and subtract that would have an answer between thirty and thirty-four, thus creating a question that teachers could later use in class.

Through planning a lesson together, each teacher had a new takeaway related to strengthening his or her own lessons on a daily basis. The team also gained a deeper understanding of the standard students needed to learn, which allowed teachers to give better feedback to students throughout the lesson. We finished the meeting talking about what the team members needed to add to their teacher edition lessons to account for each element of strong lesson design.

create the story of each lesson. If you teach a grade above or below the samples, remember to ask yourself, How does the lesson inform me of student work as part of a progression for student learning?

> **TEAM RECOMMENDATION**
>
> ## Lesson Closure for Evidence of Student Learning
>
> - As a team, commit to embedding student-led closure into your daily lessons.
>
> - Take your current activities and strategies used for closure and examine how you could make them more student centered with formative feedback to the students.
>
> - As a team, consider how you can use student-led closure to gather evidence of your students' demonstration of proficiency on the daily learning target. How can you and your team use this evidence to prepare a lesson for the next day or unit?

The last part of the template and the way you reflect on the evidence of effectiveness of the lesson set up the final chapter of this book. You will examine how you should respond to student progress during the lessons of a unit and use that input in your planning decisions for the next lesson, next unit, or sometimes the next year. You closely examine Tier 1 interventions—the intervention you provide as part of instruction as you ask, "What do I need to do differently to help every child learn?"

In the next chapter, you learn how to respond to student progress with high-quality Tier 1 mathematics intervention.

CHAPTER 12

Responding to Lesson Progress With High-Quality Tier 1 Mathematics Intervention

> In the end, all learners need your energy, your heart, and your mind. They have that in common because they are young humans. How they need you, however, differs. Unless we understand and respond to those differences, we fail many learners.
>
> —Carol Ann Tomlinson

Douglas Fisher, Nancy Frey, and Carol Rothenberg (2011) suggest that "intervention is an element of good teaching" (p. 2). That intervention begins in the classroom with the effective instruction you plan every day.

If you have been working to transform your school in pursuit of a PLC culture, then you are probably aware of three-tiered system of intervention for your students. In *Learning by Doing*, DuFour et al. (2016) describe multitiered response to intervention (RTI) as a pyramid and explain that at the base of the pyramid is Tier 1, representing the school's core instructional program as you provide your students access to both the essential mathematics grade-level learning standards and the most effective instructional strategies (for example, the elements of the instructional design framework in this book).

A key element to Tier 1 intervention in mathematics is to *look in the mirror*. Think about the core of your instructional decision-making and planning process. Did the elements of instruction that you placed into the lesson help each and every student to learn the content by the end of the lesson? You are the first line of defense for struggling mathematics students. This includes your differentiated response to learning as you provide your students and student teams with scaffolding prompts *during the lesson* to help them think of other ways to solve a problem for a higher-level-cognitive-demand task. These are in-class interventions that are at the heart of any first-step RTI model designed to support learning in class during primary instruction for every student. You and your colleagues purposefully plan your differentiated response to student learning in each lesson you teach.

Planning Your In-Class Interventions

You and your colleagues can use formative assessment information—data; information on students' prior knowledge; information about students' diverse language and cultural backgrounds (beliefs, perspectives, histories, and attitudes); and so on—to address students' learning needs, especially as you give them feedback on their progress and they take action with their peers during the lesson. By using the differentiated instruction on tasks and formative assessment strategies in class during your instruction discussed in part 2, you are not making the content easier; you are making the content more accessible to more students by the time the math class ends.

Since Tier 1 intervention is an instructional shift you embrace to help all students learn, the intervention begins with making a decision to provide your students *access* to rich mathematical higher-level-cognitive-demand tasks that will require significant differentiation during class.

Bill Barnes explains how his teacher teams began to work together to re-think this issue in his school district.

As you and your team consider how you will respond with in-class Tier 1 intervention, use table 12.1 and the teacher reflection to consider the responses to learning you can take and the potential impact on students.

Personal Story 66 BILL BARNES

When we began working with our mathematics teams to build an understanding of lower- and higher-level mathematical tasks, we found that a common teacher misconception emerged. At least one teacher on each team, often more than one, would say, "If students cannot even solve these lower-level problems, how can you expect us to ask them to solve a harder problem?" The misconception centered on a belief that mathematical thinking had to be developed strictly from procedure to application . . . but never the other way around.

Our team initiated monthly professional learning that featured a fishbowl learning laboratory so that teachers could observe student responses and engagement when introduced to higher-level-cognitive-demand tasks prior to learning procedures. Over time, teachers began seeing the value in engaging students in a worthwhile mathematical task as a strategy for student engagement in deeper mathematical thinking.

Table 12.1: Teacher Team and Student Tier 1 Responses to Learning During Effective Core Instruction.

Teacher Team Response to Learning	Student Response to Learning
Create a safe space for learning.	Students feel comfortable making mistakes and learning from each other.
Create opportunities within tasks to build social skills.	Students learn how to listen to each other, ask questions, encourage each other, and constructively formalize their thinking.
Develop positive student-teacher relationships.	Students are engaged and involved in the learning process.
Collaboratively determine what each and every student will know and be able to do at the end of every unit.	Students have a clear understanding of the learning pathway for the unit and lessons.
Choose mathematical tasks that are open ended, low-floor, high-ceiling tasks to provide multiple access points.	Students can begin the task using the skills that they have access to, using a less complex approach to a more complex way to solve the problem.
Create a student-centered community of collaborative learners.	Students view their peers as resources to support their learning.
Use multiple representations, technology, and manipulatives to make sense of the mathematics.	Students see the importance of multiple representations and how to make connections and create flexible mathematical thinking.
Provide opportunities for students to take ownership for their learning.	Students feel they are partners in their learning and understand that they can take action on the feedback provided to improve their understanding.

> **TEACHER** *Reflection*
>
> Of the teacher team responses to student learning listed in table 12.1, which actions are most *absent* from your current practice? Choose two on which to focus during your planning. List some specific actions you could take to ensure you implement the in-class Tier 1 intervention strategies this year.
>
> _____
> _____
> _____
> _____
> _____
> _____
> _____

To note, powerful core instruction described by Mattos et al. (2016) includes five components.

1. Clearly articulated essential standards
2. Success criteria for mastery (proficiency)
3. Evidence-based best practices
4. Meaningful, relevant, and student-centered instruction
5. Twenty-first century skills

How do you and your colleagues know whether or not your instructional practices are effective? How do you know which to replicate in a future unit, as an intervention strategy, or during the same unit next year? Your use of common assessments during a unit and at the end of a unit informs not only which students need additional time and support to learn the essential standards, but also which instructional practices worked most effectively across the team.

Through your analysis of student performance at the end of a math lesson against each essential standard, you can consider trends in student learning, including common misconceptions, in order to better design your responses for the next lesson based on the evidence of student learning from the lesson.

Although Tier 1 intervention is about your instructional response to student learning during class (could you teach the mathematics using different strategies that may impact a greater number of students during initial instruction?), it is also about your instructional response based on data revealed as students showcase their learning on the common during- or end-of-unit assessments you prepared (as described in *Mathematics Assessment and Intervention in a PLC at Work* [Kanold, Schuhl, et al., in press]).

Of course the expectation is that before students engage in your common unit assessments, they have received timely and specific feedback from you during those differentiated formative feedback opportunities in class. It is also expected that you and your colleagues have worked collaboratively to benefit from the data you have collected during the in-class formative feedback process, as you make final instructional adjustments to support student preparation for the common unit assessment.

Analyzing Data for Tier 1 Intervention

When analyzing data from a common assessment, there are three steps you and your colleagues can take as you reflect on the data and then use those reflections to take action on possible in-class interventions to your instructional strategies to teach the mathematics standard.

1. Select an essential learning standard, and then use student work and the corresponding feedback to identify what sets proficient and exceeding students apart related to their work.
2. Next, look at the work of students who are close to proficient and begin to see if there is a common misconception or a misunderstanding to target as a catalyst for possibly moving students to proficient since you cannot always reteach the entire standard.
3. Finally, look at trends in student work from those students who are far from proficient to identify what to target to improve learning.

You can use the questions in figure 12.1 (page 110) to evaluate your current progress toward connecting your in-class instruction and adjustments (the formative assessment process work) to the during- and end-of-unit assessment expectations from one unit to the next or one year to the next.

Personal Story 〞 TIMOTHY KANOLD

As we slowly began to embrace the idea of using student performance from our assessments to impact the nature of our lesson design, I had established an expectation that every mathematics teacher team in our district would examine their data together and then share instructional strategies that seemed to be working best for their students as a way to adjust our instruction (the mathematical tasks used, strategies for teaching the standard, and our use of formative in-class differentiation for learning).

As part of this process, I was also a teacher on our AP calculus AB team. As it came time to teach the unit on volumes by cross-section standard, we examined both our local data from the assessment the year before, as well as our national data on the AP exam (it is a major topic on the essay portion of the exam). Our analysis of student performance on the standard revealed this standard was an extremely poor performer, for a few others and me. But one teacher had reasonably good results both on our local assessment and on the national test.

We were aware that a Tier 1 response would not be to go on teaching it the way we had done in the past. Even though we felt we were good teachers of mathematics, our *strategy* for teaching the standard was not very effective. We knew our professional Tier 1 response would be to try a different strategy this year. So, Chris (she was also our team leader) showed us her use of a modeling project to build the solids, and then reveal the formulas. We all used her activity, maybe a bit adjusted, but nonetheless this same visual representation, and the intervention worked. It was in fact so effective we never had a problem with student proficiency on this standard again.

Directions: Select your next common assessment. Reflect on each statement individually, and then discuss as a team several responses taking place during instruction that will prepare students to be successful on that common assessment instrument.

	Strongly Agree				Strongly Disagree
	5	4	3	2	1
1. Teachers on our team use a variety of formative assessment strategies to encourage students to use and persevere through both standard and invented strategies.					
2. Our students represent their learning in multiple ways, such as with number lines, tables, graphs, functions, equations, and other types of models.					
3. Teachers on our team demonstrate that the answer to a mathematics problem or task is not all that matters in learning; reasoning and explaining why and how are also expected.					
4. Our students improve proficiency with both concepts and procedures through experiences with higher- and lower-level-cognitive-demand tasks.					
5. Our students increase their problem-solving and reasoning skills by solving problems together; hence teachers present students with higher-level-cognitive-demand tasks without lowering the demand during instruction and require students to justify their responses and reasoning for the solution pathways to such tasks.					
Discussion notes:					

Figure 12.1: Team discussion tool—Connecting in-class Tier 1 intervention to the end-of-unit assessment.

*Visit **go.SolutionTree.com/MathematicsatWork** for a free reproducible version of this figure.*

Use the teacher reflection to connect the results of student learning to your daily instructional routines and in-class Tier 1 interventions.

TEACHER *Reflection*

How can you continue to meet the needs of each and every student this year? What intentional in-class instructional responses to student learning can you take as a team to intervene and support students who are struggling within a unit?

_____

_____

_____

_____

_____

_____

_____

It is your team's responsibility to continually evaluate the effectiveness of your core instruction and adapt instruction as needed to meet the needs of each and every student. It is your collective response to evidence of student learning that will make the most significant impact on student success.

When answering PLC critical question 3—What will be our response when students do not learn?—there are three tiers for a reason. Tier 1 is about instructional moves you can make in class to impact student learning. Tier 2 includes a more robust out-of-class response to student learning with additional student mathematics support as needed.

Mathematics Assessment and Intervention in a PLC at Work (Kanold, Schuhl, et al., in press) provides deeper guidance on your Tier 2 intervention response, when in-class instructional shifts are not sufficient for student learning of each standard.

TEAM RECOMMENDATION

Plan for Tier 1 Interventions

- As a team, be fierce about access to rigorous and engaging learning opportunities for students.

- Discuss how you will each monitor and respond to evidence or lack of evidence of student learning as part of your instructional in-class strategies.

How do your current lesson-design practices incorporate the use of the six lesson-design elements this book outlines? Use the part 2 summary section to reflect on each of these elements.

PART 2 SUMMARY

Who dares to teach, must never cease to learn.

—*John Cotton Dana*

In part 2 of this book, you addressed the fourth team action of Mathematics in a PLC at Work—use the lesson-design elements to provide formative feedback and cultivate student perseverance during instruction. You covered the following topics related to implementing the six lesson-design elements discussed in part 1.

- Using FAST formative feedback to support student perseverance
- Designing prior-knowledge warm-up activities
- Using vocabulary as part of instruction
- Implementing balanced tasks and discourse
- Using lesson closure for evidence of learning
- Responding to lesson progress with high-quality intervention

As a team, use figure P2.2 to reflect on your current practice toward implementation of each of the six lesson-design elements highlighted in this book. And take some time to openly discuss as a team your beliefs about the purpose of a mathematics lesson using the teacher reflection.

Directions: Use the following prompts to guide your team discussion about daily lesson design and delivery.

1. How do you and your team currently communicate the essential learning standards for the unit and the daily learning target for the lesson to your students? Are your students able to state the standard for the lesson in "I can" language each day?

2. How do you and your team currently use prior-knowledge mathematics tasks to begin each lesson?

3. How do you and your team provide instruction on key academic vocabulary words or mathematics symbols for the lesson?

Figure P2.2: Team discussion tool—Using the Mathematics in a PLC at Work lesson-design elements.

continued →

4. How do you and your team use lower-level and higher-level-cognitive-demand mathematical tasks in class, and how do you connect those tasks to the essential standard for each lesson?

5. How do you and your team effectively choose whole-group and small-group discourse activities for the student lesson experience? What are the challenges of using both types of discourse?

6. How do you and your team close each lesson with a student-led summary? What do your students do?

Visit **go.SolutionTree.com/MathematicsatWork** *for a free reproducible version of this figure.*

TEACHER *Reflection*

Now that you have reflected on the six elements of effective mathematics lesson design, think about strategies you currently use on a consistent basis to engage your students in the learning process while also promoting student perseverance.

As an extension, what strategies would you like to start incorporating into your lesson-design process? Specifically list the strategy or lesson-design elements you hope to continue to improve.

In the end, instruction is both informed by and feeds into your assessment work (as described further in *Mathematics Assessment and Intervention in a PLC at Work* [Kanold, Schuhl, et al., in press]). Along the way, how you ask students to practice mathematics outside of class is an equally important aspect of your work. This is the focus of *Mathematics Homework and Grading in a PLC at Work* (Kanold, Barnes, et al., in press), which provides a deep understanding of the role homework—independent student practice—plays in your professional practice and how it affects the way in which you choose to assign student grades. *Mathematics Coaching and Collaboration in a PLC at Work* (Kanold, Toncheff, et al., in press) addresses your important work leading collaborative teams within your PLC.

Epilogue

By Timothy D. Kanold

An epilogue should serve as a conclusion to what has happened to you based on your experiences in reading and using this book. An epilogue should also serve as a conclusion to your growth and change about mathematics instruction as your teaching career unfolds.

During the 1990s and early 2000s, my colleagues and I at Adlai E. Stevenson High School District 125—the birthplace of the PLC at Work model—developed the deep mathematics instructional criteria revealed in this book. Stevenson eventually became a U.S. model for other schools and districts because of the deep inspection, revision, and eventual collective team response to the quality of the common unit-by-unit assessments in mathematics for each grade level and course.

At the time, many considered the decision to allow students to embrace their mistakes during class and talk to one another in meaningful mathematics conversations and the use of mathematics lesson designs that reflected an understanding of the lesson from the students' point of view to be a groundbreaking practices. Yet once we collectively pursued this practice with certain ferocity (the elimination of rows and use of teams of four was a pretty big deal), our student performance results began to soar. The talented co-authors of this book will tell you they experienced the same results in their school or district.

As we began our 9–12 mathematics work at Stevenson and with our K–8 feeder districts, we discovered quite a bit about our teaching. To some extent, we realized just how unbalanced we were with our instruction. Remember, we were doing this work together in the early 1990s, well before the ideas of transparency in our practice and observing and learning from one another in our professional work became as popular as they are today.

Yet we also knew that if we were to become a professional *learning* community, then experimenting together and trying out effective strategies of practice needed to become our norm. We needed to learn more about the mathematics curriculum, and we needed to do it *together*. And that is when we discovered just how unbalanced we were! And how reluctant our students were to persevere through the variety of mathematics problems (tasks) we asked them to use in order to demonstrate learning of a standard.

As an example, some of us declared the standard to be learned each day, connecting each mathematics task used in the lesson to the standard, and others of us rarely mentioned the standard at all. Learning mathematics was about learning to do a bunch of problems, we thought. Some of us used mostly lower-level-cognitive-demand tasks (procedural knowledge with rote routines), and others used mostly higher-level open-ended tasks. Some of us used application problems, and others of us used none at all. As you can imagine, these wide variances in our daily decision making caused a lot of *rigor* inequity for students in the same grade level or course.

Some of us used no warm-up activities, and others used them every day as a tap into prior knowledge. Some of us used student-led closure each day, and others just ran out of time for closure. Some of us paid attention to the academic vocabulary necessary for the lesson, and others didn't think much about vocabulary as a necessary part of a *mathematics* lesson.

Some of us used checks for understanding and mostly direct instruction from the front of the room, and others checked for understanding by moving among the students and promoting a more student-*engaged* speaking and learning environment. Some of us provided deep formative feedback, and others did not. Some of us used technology every day, and others only on occasion or not at all.

And all of us experienced our students shutting down and not engaging in the expected work of the lesson, regardless of our lesson-design style. In the words of our colleague Matt Larson, our students were *not* demonstrating productive perseverance during a mathematics lesson.

At that moment, we were far from any vision of great instruction and really had no idea about each other's daily lesson design. Our teaching practice was, quite frankly, all over the map and lacked any clear direction.

There were two issues for us as teachers of mathematics. First, we were teaching the same grade level or course, and thus our wide variance in practice was causing some of the inequities in student learning as students passed from course to course and grade to grade. Second, there was no team discussion around evidence of learning to determine which of our teacher behaviors for lesson design were best for promoting student perseverance and learning.

These realizations placed us, as a group of teachers who deeply cared about mathematics, on a journey of self-correction and incredible improvement in student learning, far beyond our own expectations.

In this book, *Mathematics Instruction and Tasks in a PLC at Work*, we have tried to help you find the right balance around design elements that have a primary impact on student perseverance in class, and most likely result in retention of learning the expected mathematics standards for your grade level or course. We also want to help you and your team learn how to use those lesson-design elements more strategically each day, and how to become more confident and transparent in using those elements for your daily mathematics lessons. We hope we succeeded as you move toward greater transparency in your practice with colleagues.

On behalf of Sarah, Matt, Bill, Jessica, Mona, and myself, may your instruction and tasks journey lead to inspired student learning each and every day.

APPENDIX

Cognitive-Demand-Level Task Analysis Guide

Table A.1: Cognitive-Demand Levels of Mathematical Tasks

Lower-Level Cognitive Demand	Higher-Level Cognitive Demand
Memorization Tasks • These tasks involve reproducing previously learned facts, rules, formulae, or definitions to memory. • They cannot be solved using procedures because a procedure does not exist or because the time frame in which the task is being completed is too short to use the procedure. • They are not ambiguous; such tasks involve exact reproduction of previously seen material and what is to be reproduced is clearly and directly stated. • They have no connection to the concepts or meaning that underlie the facts, rules, formulae, or definitions being learned or reproduced.	**Procedures With Connections Tasks** • These procedures focus students' attention on the use of procedures for the purpose of developing deeper levels of understanding of mathematical concepts and ideas. • They suggest pathways to follow (explicitly or implicitly) that are broad general procedures that have close connections to underlying conceptual ideas as opposed to narrow algorithms that are opaque with respect to underlying concepts. • They usually are represented in multiple ways (for example, visual diagrams, manipulatives, symbols, or problem situations). They require some degree of cognitive effort. Although general procedures may be followed, they cannot be followed mindlessly. Students need to engage with the conceptual ideas that underlie the procedures in order to successfully complete the task and develop understanding.
Procedures Without Connections Tasks • These procedures are algorithmic. Use of the procedure is either specifically called for, or its use is evident based on prior instruction, experience, or placement of the task. • They require limited cognitive demand for successful completion. There is little ambiguity about what needs to be done and how to do it. • They have no connection to the concepts or meaning that underlie the procedure being used. • They are focused on producing correct answers rather than developing mathematical understanding. • They require no explanations or have explanations that focus solely on describing the procedure used.	**Doing Mathematics Tasks** • Doing mathematics tasks requires complex and no algorithmic thinking (for example, the task, instructions, or examples do not explicitly suggest a predictable, well-rehearsed approach or pathway). • It requires students to explore and understand the nature of mathematical concepts, processes, or relationships. • It demands self-monitoring or self-regulation of one's own cognitive processes. • It requires students to access relevant knowledge and experiences and make appropriate use of them in working through the task. • It requires students to analyze the task and actively examine task constraints that may limit possible solution strategies and solutions. • It requires considerable cognitive effort and may involve some level of anxiety for the student due to the unpredictable nature of the required solution process.

Source: Smith & Stein, 1998. Used with permission.

References and Resources

Aguirre, J., Mayfield-Ingram, K., & Martin, D. B. (2013). *The impact of identity in K–8 mathematics: Rethinking equity-based practices*. Reston, VA: National Council of Teachers of Mathematics.

Ball, D. L., Ferrini-Mundy, J., Kilpatrick, J., Milgram, R. J., Schmid, W., & Schaar, R. (2005). Reaching for common ground in K–12 mathematics education. *Notices of the AMS, 52*(9). Accessed at www.ams.org/notices/200509/comm-schmid.pdf on June 23, 2017.

Barton, M. L., & Heidema, C. (2000). *Teaching reading in mathematics*. Aurora, CO: Mid-continent Research for Education and Learning.

Beck, I. L., McKeown, M. G., & Kucan, L. (2013). *Bringing words to life: Robust vocabulary instruction* (2nd ed.). New York: Guilford Press.

Ben-Hur, M. (2006). *Concept-rich mathematics instruction: Building a strong foundation for reasoning and problem solving*. Alexandria, VA: Association for Supervision and Curriculum Development.

Boston, M. D., & Smith, M. S. (2009). Transforming secondary mathematics teaching: Increasing the cognitive demands of instructional tasks used in teachers' classrooms. *Journal for Research in Mathematics Education, 40*(2), 119–156.

Cawelti, G. (Ed.). (1995). *Handbook of research on improving student achievement*. Arlington, VA: Educational Research Service.

Civil, M., & Turner, E. (2014). Introduction. In M. Civil & E. Turner (Eds.), *The Common Core State Standards in mathematics for English learners: Grades K–8* (pp. 1–5). Alexandria, VA: TESOL Press.

The College Board. (n.d.). *Sample questions: Math—Introduction*. Accessed at https://collegereadiness.collegeboard.org/sample-questions/math on August 21, 2017.

Dale, E., & O'Rourke, J. (1986). *Vocabulary building: A process approach*. Columbus, OH: Zaner-Bloser.

Donovan, M. S., & Bransford, J. D. (Eds.). (2005). *How students learn: History, mathematics, and science in the classroom*. Washington, DC: National Academies Press.

DuFour, R. (2015). *In praise of American educators: And how they can become even better*. Bloomington, IN: Solution Tree Press.

DuFour, R., DuFour, R., & Eaker, R. (2008). *Revisiting Professional Learning Communities at Work: New insights for improving schools*. Bloomington, IN: Solution Tree Press.

DuFour, R., DuFour, R., Eaker, R., & Karhanek, G. (2010). *Raising the bar and closing the gap: Whatever it takes*. Bloomington, IN: Solution Tree Press.

DuFour, R., DuFour, R., Eaker, R., Many, T. W., & Mattos, M. (2016). *Learning by doing: A handbook for Professional Learning Communities at Work* (3rd ed.). Bloomington, IN: Solution Tree Press.

DuFour, R., & Eaker, R. (1998). *Professional Learning Communities at Work: Best practices for enhancing student achievement*. Bloomington, IN: Solution Tree Press.

Fisher, D., Frey, N., & Rothenberg, C. (2011). *Implementing RTI with English learners*. Bloomington, IN: Solution Tree Press.

Frayer, D. A., Fredrick, W. C., & Klausmeier, H. J. (1969, April). *A schema for testing the level of concept mastery* (Working Paper No. 16). Madison: Wisconsin Research and Development Center for Cognitive Learning.

Gersten, R., Taylor, M. J., Keys, T. D., Rolfhus, E., & Newman-Gonchar, R. (2014). *Summary of research on the effectiveness of math professional development approaches* (REL 2014–010). Washington, DC: U.S. Department of Education.

Hansen, A. (2015). *How to develop PLCs for singletons and small schools.* Bloomington, IN: Solution Tree Press.

Hattie, J. (2009). *Visible learning: A synthesis of over 800 meta-analyses relating to achievement.* New York: Routledge.

Hattie, J. (2012). *Visible learning for teachers: Maximizing impact on learning.* New York: Routledge.

Hattie, J., Fisher, D., & Frey, N. (2017). *Visible learning for mathematics, grades K–12: What works best to optimize student learning.* Thousand Oaks, CA: Corwin Press.

Hattie, J., & Yates, G. (2014). *Visible learning and the science of how we learn.* New York: Routledge.

Helwig, R., Rozek-Tedesco, M. A., Tindal, G., Heath, B., & Almond, P. J. (1999). Reading as an access to mathematics problem solving on multiple-choice tests for sixth-grade students. *Journal of Educational Research, 93*(2), 113–125.

Hiebert, J. S., & Grouws, D. A. (2007). The effects of classroom mathematics teaching on students' learning. In F. K. Lester, Jr. (Ed.), *Second handbook of research on mathematics teaching and learning: A project of the National Council of Teachers of Mathematics* (pp. 371–404). Charlotte, NC: Information Age.

Jackson, K., Garrison, A., Wilson, J., Gibbons, L., & Shahan, E. (2013). Exploring relationships between setting up complex tasks and opportunities to learn in concluding whole-class discussions in middle-grades mathematics instruction. *Journal for Research in Mathematics Education, 44*(4), 646–682.

Janvier, C. (Ed.). (1987). *Problems of representation in the teaching and learning of mathematics.* Hillsdale, NJ: Erlbaum.

Johnson, D. W., & Johnson, R. T. (1999). Making cooperative learning work. *Theory Into Practice, 38*(2), 67–73.

Johnson, D. W., Johnson, R. T., & Holubec, E. J. (2008). *Cooperation in the classroom* (Rev. ed.). Edina, MN: Interaction Books.

Jones, P. S., & Coxford, A. F., Jr. (1970). *A history of mathematics education in the United States and Canada* (32nd yearbook). Reston, VA: National Council of Teachers of Mathematics.

Kagan, S. (1994). *Cooperative learning.* San Clemente, CA: Kagan.

Kagan, S., & Kagan, M. (2009). *Kagan cooperative learning.* San Clemente, CA: Kagan.

Kanold, T. D., Barnes, B., Larson, M. R., Kanold-McIntyre, J., Schuhl, S., & Toncheff, M. (in press). *Mathematics homework and grading in a PLC at Work.* Bloomington, IN: Solution Tree Press.

Kanold, T. D. (Ed.), Briars, D. J., Asturias, H., Foster, D., & Gale, M. A. (2013). *Common Core mathematics in a PLC at Work, grades 6–8.* Bloomington, IN: Solution Tree Press.

Kanold, T. D., Kanold-McIntyre, J., Larson, M. R., Barnes, B., Schuhl, S., & Toncheff, M. (in press). *Mathematics instruction and tasks in a PLC at Work.* Bloomington, IN: Solution Tree Press.

Kanold, T. D., & Larson, M. R. (2012). *Common Core mathematics in a PLC at Work, leader's guide.* Bloomington, IN: Solution Tree Press.

Kanold, T. D., Schuhl, S., Larson, M. R., Barnes, B., Kanold-McIntyre, J., & Toncheff, M. (in press). *Mathematics assessment and intervention in a PLC at Work.* Bloomington, IN: Solution Tree Press.

Kanold, T. D., Toncheff, M., Larson, M. R., Barnes, B., Kanold-McIntyre, J., & Schuhl, S. (in press). *Mathematics coaching and collaboration in a PLC at Work.* Bloomington, IN: Solution Tree Press.

Kazemi, E., & Hintz, A. (2014). *Intentional talk: How to structure and lead productive mathematical discussions.* Portland, ME: Stenhouse.

Kenney, J. M., Hancewicz, E., Heuer, L., Metsisto, D., & Tuttle, C. L. (2005). *Literacy strategies for improving mathematics instruction.* Alexandria, VA: Association for Supervision and Curriculum Development.

Kilpatrick, J., Swafford, J., & Findell, B. (Eds.). (2001). *Adding it up: Helping children learn mathematics.* Washington, DC: National Academies Press.

Larson, M. R., & Kanold, T. D. (2016). *Balancing the equation: A guide to school mathematics for educators and parents.* Bloomington, IN: Solution Tree Press.

McClure, L., Woodham, L., & Borthwick, A. (2011). *Using low threshold high ceiling tasks*. Accessed at http://nrich.maths.org/7701 on August 21, 2017.

McEwan-Adkins, E. K. (2010). *40 reading intervention strategies for K–6 students: Research-based support for RTI*. Bloomington, IN: Solution Tree Press.

Miura, I. T., & Yamagishi, J. M. (2002). The development of rational number sense. In B. Litwiller & G. Bright (Eds.), *Making sense of fractions, ratios, and proportions: 2002 yearbook* (pp. 206–212). Reston, VA: National Council of Teachers of Mathematics.

Morris, A. K., Hiebert, J., & Spitzer, S. M. (2009). Mathematical knowledge for teaching in planning and evaluating instruction: What can preservice teachers learn? *Journal for Research in Mathematics Education, 40*(5), 491–529.

National Board for Professional Teaching Standards. (2010). *Mathematics standards for teachers of students ages 11–18+* (3rd ed.). Arlington, VA: Author.

National Council of Teachers of Mathematics. (1980). *An agenda for action* [Pamphlet]. Reston, VA: Author.

National Council of Teachers of Mathematics. (1991). *Professional standards for teaching mathematics*. Reston, VA: Author.

National Council of Teachers of Mathematics. (2009). *Focus in high school mathematics: Reasoning and sense making*. Reston, VA: Author.

National Council of Teachers of Mathematics. (2014). *Principles to actions: Ensuring mathematical success for all*. Reston, VA: Author.

National Council of Teachers of Mathematics. (2018). *Catalyzing change in high school mathematics: Initiating critical conversations*. Reston, VA: Author.

National Governors Association Center for Best Practices & Council of Chief State School Officers. (2010). *Common Core State Standards for mathematics*. Washington, DC: Authors. Accessed at www.corestandards.org/Math on November 27, 2017.

Pellegrino, J. W., & Hilton, M. L. (Eds.). (2012). *Education for life and work: Developing transferable knowledge and skills in the 21st century*. Washington, DC: National Academies Press.

Popham, W. J. (2011). *Transformative assessment in action: An inside look at applying the process*. Alexandria, VA: Association for Supervision and Curriculum Development.

Reeves, D. (2011). *Elements of grading: A guide to effective practice*. Bloomington, IN: Solution Tree Press.

Reeves, D. (2016). *Elements of grading: A guide to effective practice* (2nd ed.). Bloomington, IN: Solution Tree Press.

Resnick, L. B. (2006). Do the math: Cognitive demand makes a difference. *Research Points: Essential Information for Education Policy, 4*(2), 1–4. Accessed at www.aera.net/Portals/38/docs/Publications/Do%20the%20Math.pdf on July 10, 2017.

Riccomini, P. J., Smith, G. W., Hughes, E. M., & Fries, K. M. (2015). The language of mathematics: The importance of teaching and learning mathematical vocabulary. *Reading and Writing Quarterly, 31*(3), 235–252. Accessed at www.tandfonline.com/doi/full/10.1080/10573569.2015.1030995?src=recsy on August 21, 2017.

Rubenstein, R. (2007). Focused strategies for middle-grades mathematics vocabulary development. *Mathematics Teaching in the Middle School, 13*(4), 200–207.

Rubenstein, R., & Thompson, D. R. (2002). Understanding and supporting children's mathematical vocabulary development. *Teaching Children Mathematics, 9*(2), 107–112.

Schimmer, T. (2016). *Grading from the inside out: Bringing accuracy to student assessment through a standards-based mindset*. Bloomington, IN: Solution Tree Press.

Schmoker, M. (2011). *FOCUS: Elevating the essentials to radically improve student learning*. Alexandria, VA: Association for Supervision and Curriculum Development.

Schoenbach, R., Greenleaf, C., Cziko, C., & Hurwitz, L. (1999). *Reading for understanding: A guide to improving reading in middle and high school classrooms*. San Francisco: Jossey-Bass.

Silver, E. (2010). Examining what teachers do when they display their best practice: Teaching mathematics for understanding. *Journal of Mathematics Education at Teachers College, 1*(1), 1–6.

Slavin, R. E. (2014). Making cooperative learning powerful. *Educational Leadership, 72*(2), 22–26.

Smith, M. S., Steele, M. D., & Raith, M. L. (2017). *Taking action: Implementing effective teaching practices in grades 6–8*. Reston, VA: National Council of Teachers of Mathematics.

Smith, M. S., & Stein, M. K. (1998). Selecting and creating mathematical tasks: From research to practice. *Mathematics Teaching in the Middle School, 3*(5), 344–350.

Smith, M. S., & Stein, M. K. (2011). *5 practices for orchestrating productive mathematics discussions*. Reston, VA: National Council of Teachers of Mathematics.

Stein, M. K., Remillard, J., & Smith, M. S. (2007). How curriculum influences student learning. In F. K. Lester, Jr. (Ed.), *Second handbook of research on mathematics teaching and learning: A project of the National Council of Teachers of Mathematics* (pp. 319–370). Charlotte, NC: Information Age.

Stigler, J. W., & Hiebert, J. (1999). *The teaching gap: Best ideas from the world's teachers for improving education in the classroom*. New York: Free Press.

Tyson, K. (2013, May 26). *No tears for tiers: Common Core tiered vocabulary made simple* [Blog post]. Accessed at www.learningunlimitedllc.com/2013/05/tiered-vocabulary on July 11, 2017.

Vygotsky, L. S. (1978). Interaction between learning and development. In M. Gauvain & M. Cole (Eds.), *Readings on the development of children* (pp. 34–40). New York: Scientific American Books.

Wiliam, D. (2011). *Embedded formative assessment*. Bloomington, IN: Solution Tree Press.

Wiliam, D. (2018). *Embedded formative assessment* (2nd ed.). Bloomington, IN: Solution Tree Press.

Zike, D. (2003). *Dinah Zike's teaching mathematics with foldables*. Columbus, OH: Glencoe/McGraw-Hill.

Index

A

academic language
  defined, 27
  discussion tools, 29, 73–75, 76
  importance of, 27, 71
  incorporating, 28–31
  modeling, 30
  strategies and challenges, 73
  tiers of, 28
  when and how to teach, 72–76
accurate feedback, 63
Adding It Up (NCR), 51
advancing prompts, 87, 88
Aguirre, J., 8
Alibali, M. W., 19
assessing prompts, 87, 88
attention and inattention, student, 40

B

Balancing the Equation (Larson and Kanold), 36
Barnes, B., 108
Barton, M. L., 28
Beck, I. L., 28
Ben-Hur, M., 27
Boston, M., 33–34

C

cascading inattention, 40
closure. *See* lesson closure
cognitive-demand tasks, lower- and higher-level
  analysis guide, 117
  choosing, 33–38, 69
  defined, 34
  discussion tool, 37
  examples of, 34, 35, 78–81, 117
  implementing, 77–82
  importance of, 33, 34
  rigor-balance of, 34
Conant, F. R., 99
Coxford, A. F., 33

D

Dana, J. C., 113
discourse, whole- and small-group
  defined, 39
  discussion tools, 44, 85, 86, 88, 93, 94–95
  facilitating and balancing, 39–45
  importance of, 40
  managing, 90–98
  questioning techniques, 83–88
  reflection questions, 40
  rights and norms, 90
  seating charts, 91–92
  small-group applications, 42–43, 69, 86–88
  student self-evaluation form, 95, 96–97
  whole-group applications, 41, 82–86
DuFour, R., 1, 107

E

Eaker, R., 1, 107
Education for Life and Work: Developing Transferable Knowledge and Skills in the 21st Century (Pellegrino and Hilton), 34
Einstein, A., 67

Engle, R. A., 99
equity, professional learning communities and, 1–2
essential learning standards. *See* learning standards
evidence of learning. *See* lesson closure

F

fair feedback, 63
FAST feedback, 62–63, 87
feedback
    accurate (effective), 63
    fair, 63
    FAST, 62–63, 87
    specific, 63
    timely, 63
Fisher, D., 30, 72, 107
formative assessment
    checking for understanding versus, 63–66
    discussion tool, 89
    implementing, 77–82
    monitoring actions and results, 88–90
Franklin, B., 39
Frey, N., 30, 72, 107

H

Hansen, A., 3
Hattie, J., 13, 30, 40, 47, 63, 72, 97, 99
Heidema, C., 28
Helwig, R., 30
higher-level-cognitive demand tasks. *See* cognitive-demand tasks, lower- and higher-level
Hilton, M. L., 34

I

I can statements, 14, 15–17, 78
interventions. *See* Tier 1 intervention

J

Jones, P. S., 33

K

Kanold, T. D., 14, 33, 36, 42, 48, 65, 82, 92, 101–102, 110, 115
Kanold-McIntyre, J., 30, 42, 43, 72, 97
Kenney, J., 28
King, M. L., Jr., 7
Kucan, L., 28

L

Larson, M. R., 33, 36, 54, 116

Learning by Doing (DuFour, Eaker, Dufour, Many, and Mattos), 107
learning standards
    clarity and identifying, 13–18
    unpacking, 14
learning targets, 14, 15–18, /91–92
lesson closure
    activities, facilitating, 99–101
    closing prompts, examples of, 49
    defined, 47
    discussion tools, 48, 100–101
    evaluation of the effectiveness of, 102–106
    student-led summaries, 47
lesson design
    See also under each element
    discussion tools, 9, 10, 55, 113–114
    evaluation tools, 10, 11, 103–104
    elements of, 10
    planning, 51, 58–59
    purpose of, 10
    tools, 9, 10, 11, 52–53
lower-level-cognitive demand tasks. *See* cognitive-demand tasks, lower- and higher-level

M

Many, T. W., 107
Martin, D. B., 8
mathematical literacy, 36
mathematical tasks
    See also tasks, lower- and higher-level-cognitive demand
    defined, 33–34
mathematics instruction
    purpose of, 7–8
    relevant versus meaningful, 8
Mattos, M., 107, 109
Mayfield-Ingram, K., 8
McKeown, M. G., 28
meaningful mathematics, 8
Meisels, S. J., 61
Miura, I., 71

N

National Board for Professional Teaching Standards, 40
National Council of Teachers of Mathematics (NCTM), 40, 47, 51, 77, 83
National Research Council, 51

P

Pellegrino, J. W., 34
Popham, W. J., 63
Principles to Actions (NCTM), 83
prior knowledge
    discussion tool, 68
    guidelines to consider, 69–70
    importance of, 19
    task-planning tool, 24, 25
    warm-up activities (tasks), choosing, 19–26
procedural fluency, 34, 36
professional learning communities (PLCs)
    critical questions, 2, 4, 51
    equity and, 1–2
professional learning communities framework, mathematics in, 3, 4, 8–11
Professional Learning Communities at Work® (PLC) process, role of, 1–2

Q

questioning techniques, 83–88

R

Raith, M. L., 39
Reeves, D., 63
reflect, refine, and act cycle, 2–3
relevant mathematics, 8
response to intervention (RTI). *See* Tier 1 intervention
Rothenberg, C., 107
Rubenstein, R. N., 27, 28, 71

S

scaffolding prompts, 87
Schmoker, M., 41
Schuhl, S., 19–20, 35, 102, 105
Sidney, P. G., 19
small-group discourse. *See* discourse, whole- and small-group
Smith, M., 33–34, 39, 89
specific feedback, 63
Steele, M., 39
Stein, M. K., 34, 89
student-led summaries, 47
student perseverance, 19, 40, 62–66, 77, 82–88

T

tasks, lower- and higher-level-cognitive demand. *See* cognitive-demand tasks, lower- and higher-level
Thompson, D. R., 27, 28

Tier 1 intervention
    analyzing data for, 109–111
    discussion tool, 110
    planning, 107–109
    responses to learning, 108
    role of, 107
timely feedback, 63
Tomlinson, C. A., 107
Toncheff, M., 100
Tyson, K., 28

U

understanding, checking for, 63–66
    See also lesson closure
unstucking prompts, 87

V

Visible Learning for Mathematics (Hattie, Fisher, and Frey), 72, 99
vocabulary. *See* academic language
Vygotsky, L., 86

W

warm-up activities. *See* prior knowledge
whole-group discourse. *See* discourse, whole- and small-group
Wiliam, D., 62

Y

Yamagishi, J., 71
Yates, G., 40

Every Student Can Learn Mathematics series
Timothy D. Kanold et al.
Discover a comprehensive PLC at Work® approach to achieving mathematics success in K–12 classrooms. Each book offers two teacher team or coaching actions that empower teams to reflect on and refine current practices and routines based on high-quality, research-affirmed criteria.
BKF823   BKF824   BKF825   BKF826

Mathematics Assessment and Intervention in a PLC at Work®
*Timothy D. Kanold, Sarah Schuhl, Matthew R. Larson,
Bill Barnes, Jessica Kanold-McIntyre, and Mona Toncheff*
Harness the power of assessment to inspire mathematics learning. This user-friendly resource shows how to develop high-quality common assessments, and effectively use the assessments for formative learning and intervention. The book features unit samples for learning standards, sample unit exams, and more.
BKF823

Mathematics Homework and Grading in a PLC at Work®
*Timothy D. Kanold, Bill Barnes, Matthew R. Larson,
Jessica Kanold-McIntyre, Sarah Schuhl, and Mona Toncheff*
Rely on this user-friendly resource to help you create common independent practice assignments and equitable grading practices that boost student achievement in mathematics. The book features teacher team tools and activities to inspire student achievement and perseverance.
BKF825

Mathematics Coaching and Collaboration in a PLC at Work®
*Timothy D. Kanold, Mona Toncheff, Matthew R. Larson,
Bill Barnes, Jessica Kanold-McIntyre, and Sarah Schuhl*
Build a mathematics teaching community that promotes learning for K–12 educators and students. This user-friendly resource will help you coach highly effective teams within your PLC and then show you how to utilize collaboration and lesson-design elements for team reflection, data analysis, and action.
BKF826

Solution Tree | Press a division of Solution Tree

Visit SolutionTree.com or call 800.733.6786 to order.

GLOBAL PD

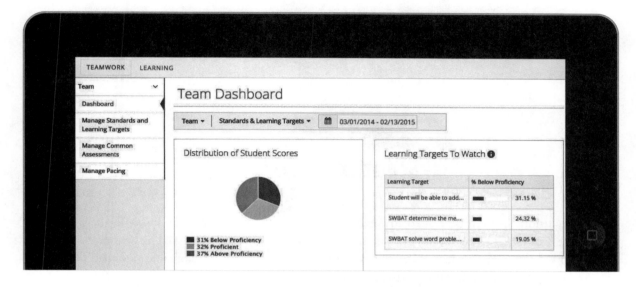

The **Power to Improve**
Is in Your Hands

Global PD gives educators focused and goals-oriented training from top experts. You can rely on this innovative online tool to improve instruction in every classroom.

- Get unlimited, on-demand access to guided video and book content from top Solution Tree authors.

- Improve practices with personalized virtual coaching from PLC-certified trainers.

- Customize learning based on skill level and time commitments.

▶ **REQUEST A FREE DEMO TODAY**
SolutionTree.com/GlobalPD

Solution Tree's mission is to advance the work of our authors. By working with the best researchers and educators worldwide, we strive to be the premier provider of innovative publishing, in-demand events, and inspired professional development designed to transform education to ensure that all students learn.

The National Council of Teachers of Mathematics is a public voice of mathematics education, supporting teachers to ensure equitable mathematics learning of the highest quality for all students through vision, leadership, professional development, and research.